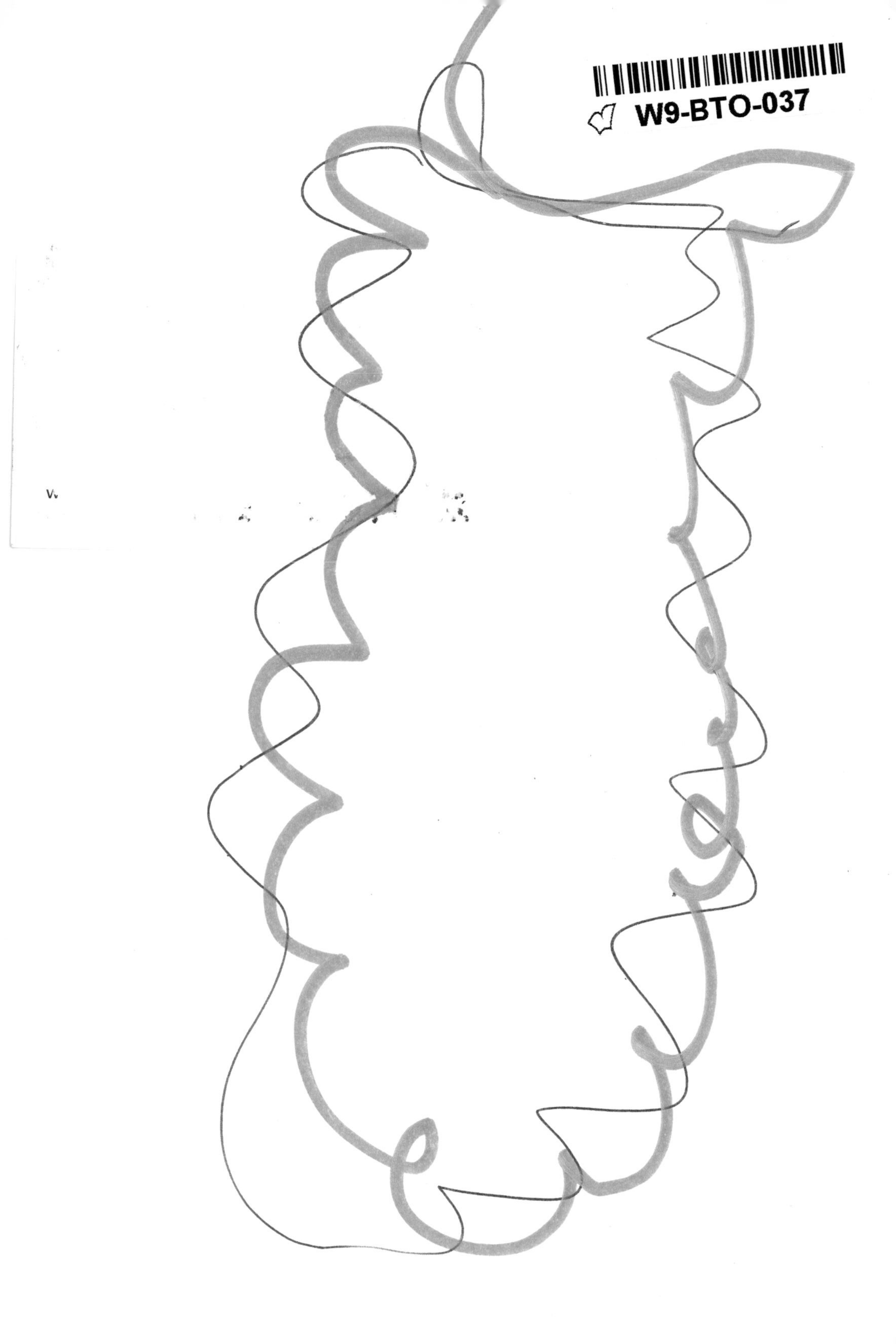

second nature

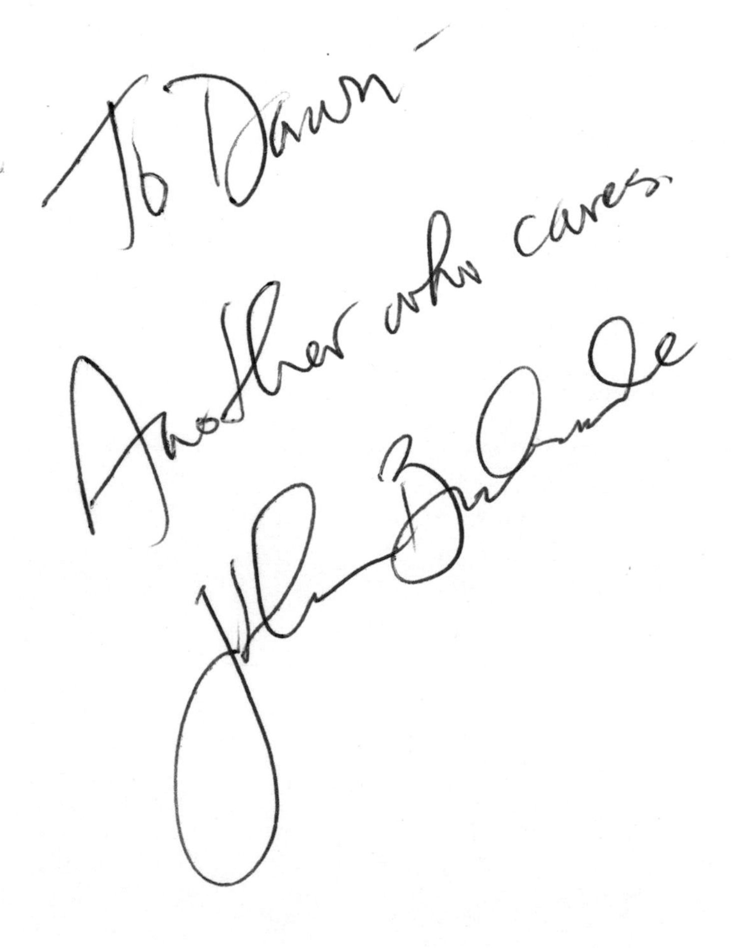

Also by Jonathan Balcombe and Palgrave Macmillan:

Pleasurable Kingdom:
Animals and the Nature of Feeling Good

second nature

the inner lives of animals

Jonathan Balcombe

Foreword by J. M. Coetzee

SECOND NATURE

First published in 2010 by
PALGRAVE MACMILLAN®
in the United States—a division of St. Martin's Press LLC,
175 Fifth Avenue, New York, NY 10010.

Where this book is distributed in the UK, Europe and the rest of the world, this is by Palgrave Macmillan, a division of Macmillan Publishers Limited, registered in England, company number 785998, of Houndmills, Basingstoke, Hampshire RG21 6XS.

Palgrave Macmillan is the global academic imprint of the above companies and has companies and representatives throughout the world.

Palgrave® and Macmillan® are registered trademarks in the United States, the United Kingdom, Europe and other countries.

ISBN: 978–0–230–61362–1

Library of Congress Cataloging-in-Publication Data

Balcombe, Jonathan P.
Second nature : the inner lives of animals / Jonathan Balcombe.
p. cm.
Includes bibliographical references and index.
ISBN 978–0–230–61362–1
1. Animal behavior—Anecdotes. 2. Animal intelligence—Anecdotes. 3. Animal psychology—Anecdotes. 4. Social behavior in animals—Anecdotes. I. Title.

QL791.B26 2009
591.5—dc22 2009030770

A catalogue record of the book is available from the British Library.

Design by Newgen Imaging Systems (P) Ltd., Chennai, India.

First edition: March 2010

10 9 8 7 6 5 4 3 2 1

Printed in the United States of America.

To hyenas, spiders, bats, snakes, whip scorpions,
and all the other beings deemed foul and loathsome—this one's for you

contents

Foreword by J. M. Coetzee ix

Acknowledgments xiii

Part I Experience

One Introduction 7
Two Tuning In: Animal Sensitivity 15
Three Getting It: Intelligence 31
Four With Feeling: Emotions 45
Five Knowing It: Awareness 61

Part II Coexistence

Six Communicating 83
Seven Getting Along: Sociability 103
Eight Being Nice: Virtue 121

Part III Emergence

Nine Rethinking Cruel Nature 143
Ten Homo Fallible 163
Eleven The New Humanity 185

Notes 205

Index 231

be able to spell out what constitutes a scientific proof of the kind they believe in. In the area of animal physiology, criteria of proof usually come framed in statistical terms; the statistics in turn depend on the mathematics of probability, and the mathematics of probability rests on rarefied philosophical assumptions. All in all, a body of difficult theory which even the professional scientific practitioner revisits only rarely and more or less takes on faith.

People believe what they want to believe; people believe in scientific proof without quite knowing what scientific proof is. Nevertheless, over much of the world, science—Western science—has acquired such prestige that people in general will be too embarrassed to deny what science claims: that any account of how things work in this world of ours cannot really be "true" (that is to say, valid) until it has been endorsed by science, or conversely that no account can really be true if science has shown it to be untrue. Such is the prestige of science that we can say it has taken over the authority that religion used to have.

Nevertheless, while not questioning the authority of science, ordinary people have no difficulty in holding unscientific beliefs at the same time, even when such beliefs turn out to be at odds with scientific truth, "the facts of science." This ambivalence, often surreptitious, constitutes a critique of science, and scientific standards of truth, though not one that is expressed in a reasoned, articulate way, the way in which a scientist might speak. Most commonly the critique is tacit, as, for instance, when a patient takes the medicine her (Western, scientific) doctor has prescribed while without telling her doctor taking "alternative" medicines too. The medical science you learn in medical school, she is in effect saying, is only one way of understanding and treating disease, and not necessarily the best.

Second Nature is the work of a scientist, but clearly not of a scientist from the animal-science establishment. It is a book notable for—to use a paradoxical term—the humanity with which it approaches the lives of animals. Jonathan Balcombe is, philosophically speaking, a Darwinian, as all biologists nowadays tend to be. Nevertheless, he is prepared to reflect on instances of animal behavior of the kind that orthodox Darwinians find hard to fit into their scheme, and sometimes to concede that such behavior exhibits what it seems to exhibit: compassion, for instance, or selflessness. In other words, he is prepared to give animals the benefit of the doubt. Why should it always be the

foreword

It used to be thought—and probably still is, in some quarters—that what set man apart from mere beasts was the possession of reason. The argument was a subtle one, with profound implications. Reason—God-given reason—was what the mind of man had in common with the mind of God. It was only because his mind was like (even if infinitely inferior to) his creator's that man was able to comprehend, to however minuscule an extent, how the world worked. Mere animals might be able to respond and adjust to the world in which they found themselves, but they would never, properly speaking, be able to understand it because their minds lacked the active principle infusing the universe, namely reason. They (together with their minds) would always be merely part of nature; they could never be masters of nature.

From the point of view of animals, it is one of the darker ironies of history that the role of being the expert authority on them has fallen not to (say) husbandmen or hunters, but to scientists, the ultimate practitioners of (human) reason and therefore, in a sense, their hereditary enemies. Do fish feel pain? Can parrots think? For an authoritative answer, a respectable answer, an answer we can believe in, we must resort to science: to the expert on the piscine nervous system, the expert on the avian cortex.

Ordinary people do not need to have something proved to them scientifically before they will believe it. They believe it because their parents believed it, or because it is accepted as so in the circles in which they move, or because figures of authority say it is so. Mostly, however, people believe what they want to believe, what it suits them to believe. Thus: fish feel no pain.

Only a tiny minority are prepared to believe only what has been proved to be scientifically true, and of that minority only a fraction will doubters who get the benefit of the doubt, he in effect asks? Thus: Why should the onus fall on animals, species by species, to prove they are sentient? Why should the burden of proof not fall on science to demonstrate they are not?

Balcombe is not afraid to attribute to some species of animals "higher" moral feelings like gratitude, and complex if nebulous states of awareness like a sense of mortality. He is even prepared to entertain the notably un-Darwinian idea that virtue—doing something selfless for no tangible benefit to oneself—may be its own reward.

For a scientist, he is remarkably open to what we can call the appeal of intuitive, trans-species fellow-feeling, an appeal that some of his colleagues would dismiss as projection. He takes seriously—not just as rhetoric—William Blake's immortal question, "How do you know but that every bird that cleaves the aerial way is not an immense world of delight closed to your senses five?"

The most interesting parts of his book concern the inner lives of animals and the ways in which sensory experience, feeling, emotion, and consciousness may be seen to interact in different species. While cautious about our ability to inhabit the minds of other species, he nonetheless argues that, by dint of attentive observation of the everyday activities of animals, particularly those that are "like" us, we can to a degree come to see the world—our common world—through their eyes and thus to a degree experience, vicariously, "their" world.

There are plenty of stories, going back to Aesop and beyond, about the cunning, the intelligence, and even the wisdom of animals—what we might call folk evidence attesting to their mental capacities. It is part of the ethos of science to maintain a healthy skepticism toward anecdotal evidence of this kind (one of the few things science does not turn a skeptical eye on is its own ethos of skepticism). Yet at a certain point, Balcombe suggests, principled skepticism can turn into dogged foot-dragging: here he points to a tendency among investigators, each time animals pass the tests we have set for them, to raise the bar slightly higher.

Ever since Aristotle's time we have made the possession of intelligence—intelligence of the kind that enables one to construct intricate machines or ingenious philosophical theories—the crucial test, the test that distinguishes higher from lower, man from beast. Yet why should the crucial test not be a quite different one: for instance, the possession of a faculty that enables a being to find its way home over long distances? Is the explanation perhaps that the latter is one that *Homo sapiens* would find it hard to pass?

Balcombe is particularly forthright in his criticism of the *idée reçue,* fostered by such early ethologists as Robert Ardrey, that nature must always be red in tooth and claw, a site of relentless struggle, of kill or be killed. Far from being absorbed in a grim battle to survive, he contends, animals actually enjoy life minute by minute, day by day. The very extravagance of certain evolutionary developments, like the peacock's tail, is proof that life is not parsimonious but has resources to squander. He has harsh words for television producers who, to satisfy the human appetite for blood, favor scenes of cruelty in nature documentaries, as well as for such intellectual luminaries as Richard Dawkins and Daniel Dennett, behind whose delight in emphasizing the misery and destruction to be found in nature, he suggests, lies an undeclared motive: to excuse mankind for its cruel treatment of other species.

As for vivisection, he has, he confesses, a distaste for the practice going back to his student days. Although he does not ask the truly radical antivivisection question—Can the suffering and death of animals in laboratories ever be justified, even if it leads to advances in scientific knowledge?—he is critical of the treatment routinely meted out to animals in research laboratories. "When I read [certain] research protocols," he writes, "I find myself shaking my head in wonderment that my species can methodically poison sentient animals, observing with clinical detachment and taking meticulous notes and measurements as they slowly die."

All in all, Balcombe is a rare being, a scientist who has escaped the narrow orthodoxies of institutional science, an intelligent human being who is more than ready to recognize intelligences of other kinds, an intuitive and empathetic observer who nevertheless does not abandon the highest standards of intellectual inquiry.

—J. M. Coetzee

acknowledgments

Marilyn and Emily were a constant support throughout the long and sometimes taxing process of cultivating this book. Thank you for standing up for readers everywhere when my ambitions ran astray. To Megan and Mica, thank you for desiring a warm lap, and for naïvely thinking that you are the primary beneficiaries of a good belly-rub.

To my close friends Ken Shapiro and Martin Stephens, who read major portions of the manuscript and provided sage advice, thank you for all your suggestions that I heeded, and I take full responsibility for those few that I didn't. To my editor, Luba Ostashevsky, thank you for being a cheerful voice of encouragement, and for helping turn a manuscript into a book worthy of its cover. To my agent, Sheila Ableman, thank you for being a tireless and effective advocate for my work. To other members of the Palgrave team, most notably Laura Lancaster, Mark LaFlaur, and Erica Warren, thank you for your contributions to this project.

My gratitude also to the following for your knowledge and creative input: Philippe Aghion, Frank Ascione, Maureen Balcombe, Neal Barnard, Gay Bradshaw, Culum Brown, Herman Daly, Jennifer Fearing, Hope Ferdowsian, Amy Fitzgerald, Barbara French, William Gilly, Pauliina Laurila, Rebecca Lewiston, Lori Marino, Jason Matheny, Ian McCallum, Cheryl Miller, Laura Moretti, David Reed, Paul Shapiro, Con Slobodchikoff, Patrick Sullivan, Hanna Tuomisto, Paul Turner, Fransje van Riel, and Steve Wang.

To Steve Mandel, Connie Pugh, Fransje van Riel, and Mike Howell who were generous in contributing their photographic talents to the project.

To all those unnamed whose ideas have informed and helped shape my own over the years, thank you. And to the countless, anonymous two-, four-, six- or more-legged and finned beasts sharing the journey—thank you for your infinite capacity to surprise and delight.

Finally, a special thanks to my parents, Maureen and Gerry, for sharing in my fascination for animals from the beginning, and for cultivating curiosity and compassion.

Part I

Experience

In 2004 a study in the journal *Human Nature* tested the aesthetic preferences of chickens versus those of humans.[1] On first blush, it might sound silly to attribute to chickens a capacity for aesthetic taste that we associate with humans. But there it was, a group of chickens who were trained to make choices by pressing a button with their beaks, lined up in a laboratory presented with digitized photos of thirty-five young men and women. In another room were seven female undergraduates instructed to choose the most attractive male face, and seven male students who were to choose the most attractive female face. When the chickens cast their votes, their preferences were almost identical to those of the students. The preference overlap was an uncanny 98 percent.

What are we to make of this? The authors, from Stockholm University, designed the study to better understand what biological forces shape people's idea of beauty. They wanted to know if there is a template of beauty hardwired into our imagination that drives our sexual choices. One way to test the hypothesis is to see if you can find the same instinct in a species that, from the human perspective, ranks lower on the evolutionary totem pole.

As an ethologist, a biologist specializing in animal behavior (some ethologists study human character and behavior), I am particularly fascinated by the other side of the problem. What might this study reveal about how chickens perceive the world? Do chickens find human faces attractive in the same way that we do? I'm unaware of any follow-up studies to try to unravel that particular mystery, but it seems doubtful that the chickens' preferences reflect an aesthetic attraction to a given human, per se. Other studies have suggested that bilateral symmetry, or the physical balance of features, is subconsciously valued by humans in rating faces, and perhaps it is also symmetry that the chickens are responding to. Whatever the basis for the chickens' choices, they evince an acute ability to discern on par with humans. That they are discerning a different species is even more impressive.

For most of the twentieth century it was considered erroneous to entertain questions about how animals might feel and what they may be thinking. However, since the 1970s scientists have been showing a growing interest in animals' hearts and minds. Some surprising new information has been surfacing. This book will discuss the amazing discoveries that naturalists are making about animal behavior. For example, starlings can become optimistic or pessimistic depending on their living circumstances, fishes choose trustworthy partners for dangerous predator inspection missions, deer mice are control freaks, and prairie dogs have a special call for a human with a gun. As for chickens, the common perception that they are barnyard dunces is more a betrayal than a portrayal. As we'll see in the chapters ahead, chickens are autonomous individuals with mental and emotional experiences, and lives worth living. They communicate with a nuanced call repertoire that refers to specific aspects of their surroundings. They may utter these calls honestly, and they may also use them deceptively.

There is, by the way, a certain poignancy to the question of how chickens may perceive humans. Why should chickens find us beautiful at all? Today, in the United States alone we kill and cause suffering to more chickens than there are human beings on the entire planet.

My chief aim in this book is to close the gap between human beings and animals—by helping us understand the animal experience, and by elevating animals from their lowly status. Part I, "Experience," examines a wealth of evidence now emerging that animals are more perceptive, intelligent, aware, and emotional than humans usually give them credit for. Part II, "Coexistence," focuses on animals' interactive natures: their sophisticated modes of communication, their sociability, and their virtues. Just thirty years ago it was scientific heresy to ascribe such emotions as delight, boredom, or joy to a nonhuman. It was unheard of to say that fishes feel pain, never mind that they have culture, and it would have been a joke to entertain the idea that animals might actually have some moral awareness. As we'll see in the pages that follow, researchers around the world have found that there is more thought and feeling in animals than humans have ever imagined.

By showing that animals think and feel richly, that they are highly sentient and sometimes even virtuous, I hope to convince you that we cannot continue to treat animals cruelly or carelessly. Part III, "Emergence," focuses on the evolving, troubled relationship between humans and animals. With all that we have learned about animals, we can no longer plead ignorance. Science is now revealing that animals are more aware and sophisticated than we thought, proving that the popular portrayals and perceptions of wild nature are biased and impoverished. In Chapter 9, I'll address the perception of nature as strictly harsh and selfish, with examples of the benignity and sustainability with which animals conduct their affairs. In Chapter 10, I put human nature under the microscope, arguing that in light of our ongoing history of violent conflict and institutionalized indifference it is hypocritical to characterize animal life as uncivilized. I close with my view of how we should turn our relationship with animals in a kinder direction. A new humanity is in order—one that demands a new ethic of mutual tolerance and respect for the other creatures doing their best to share the planet with us.

Today these ideas are finally being given the attention they have long deserved. We humans have been over-absorbed in the destiny of our own species, and one of the lessons to be learned from climate change, biodiversity loss, urban sprawl, ethnic conflicts, and economic downturns is that when we abuse or neglect fauna and flora, we also harm ourselves in the process.

CHAPTER ONE

introduction

I sense that, without sensitivity to physical pain and pleasure, men... would not have known self-interest;... and consequently no just or unjust acts; thus physical sensitivity and self-interest were the authors of all justice.
—Claude-Adrien Helvétius, *De l'esprit* ("On the Mind"), 1758[1]

Becoming a Biologist

I've been fascinated by animals for as long as I remember. According to my mother, trips to London's Whipsnade Zoo were among the highlights of my early childhood. We immigrated to New Zealand when I was three, and I recall petting semi-tame kangaroos at a park in Australia while en route. By the time my family arrived in Toronto, Canada, in 1967, I was a dinosaur-starved boy of eight. Seeing them for the first time at the Royal Ontario Museum thrilled me, as they have delighted countless others.

By the time I finished high school, my affinity for animals had grown, and I enrolled at Toronto's York University to study biology. I learned soon enough that studying animals at this level was often not in the animals' best interests. The introductory biology labs included a parade of formaldehyde-preserved specimens ready for pinning and flaying in rubber-bottomed dissection trays. I remember one midterm exam in which each student was handed a large, freshly killed bull-frog and instructed to dissect and label a set of prescribed body parts. I looked at the limp, shiny form in front of me and was saddened that her life was taken away for such a paltry reason. The abdominal cavity revealed a dense mass of eggs—several hundred nascent tadpoles.

A year later—still stinging from a lab in which our instructor had callously snipped off the legs and wings of a live male locust with a pair of dissecting scissors—I performed a small act of animal liberation. We had crossbred fruit flies (*Drosophila melanogaster*) bearing different phenotypes, and it was time to record the distribution of characteristics in the next generation. Flies were kept in small plastic vials with a ball of cotton and a pasty food medium whose odor permeated the classroom. Counting the number of flies with white versus red eyes required exposing them to ether to immobilize them. The flies were then sprinkled onto a piece of white paper to be examined and counted. When the data collection was complete, the flies had no further use to genetics, and our instructions were to tip them into a small glass dish of oil placed at the center of each desk. The "morgue," as it was called, was to be the diminutive *Drosophilas*' final resting place.

Rocking the boat never came easily to me, but I was having none of this. Once my little pile of dipterans had been counted, I pushed them off the edge of the paper where they were camouflaged against the

The author holding a hognose snake caught and released while he was studying bats in Texas. (Photo by Stephen Killefer.)

black desktop. As we recorded our data, I kept one eye on them. The ember of life soon rekindled, and within minutes the pile was twitching and humming as tiny legs and wings beat their way out of the ether fog. They staggered onto their spindly legs before regaining their senses and launching forth. I was thrilled as they took flight.

The flies were my first step in refusing to conduct scientific research that treated nonhuman life as dispensable. They also charted a path for the values I want to bring to the study of animals. As I became more aware of institutionalized abuses of animals, I identified a niche for my future: animal protection.

Bat Years

Bats are fabulously diverse. If you lined up each kind of mammal living on earth today, every fourth one would be a bat. There are more than a thousand species, which ranks them second only to rodents in diversity among the twenty-nine living mammal groups. Bats owe much of their success to the important evolutionary breakthrough of self-powered flight. Combined with the evolution of a hi-fi sonar system for orienting and foraging in the dark, flight allowed bats to muscle in on some prosperous ecological niches, notably a banquet of fruit and nocturnal insects.

One of the rewards of studying bats has been to help dissolve their demonized reputation. Mexican free-tailed bats (*Tadarida brasiliensis mexicana*), on which I wrote my PhD dissertation, are renowned for their vast maternity colonies numbering up to twenty million. Pregnant females migrate north each spring from their Mexican wintering range and settle in several limestone caverns scattered mainly across Texas and New Mexico. In early June, each bat gives birth to a single pup. At this time, 90 percent of the species' population of females and young are restricted to perhaps a half dozen caves. My field research in southern Texas was part of years of observations and experiments by a team of biologists aimed at understanding the free-tails' reunion behavior. Mother bats maintain their naked newborn pups in dense crèches of hundreds of thousands, even millions, of other pups on the walls of the pitch-dark caverns. We called the baby bats "pinkies" because of their color, and because each is about the size of your little finger. The newborns are packed so densely on the convoluted cave walls that 150 of them would fit into the space of this page. Mothers leave their pups on the crèche to be able to forage or rest, and they return to nurse them several times a day.

How these reunions take place in these dense, dark, cacophonous caves is a case study in animals' perceptual abilities. Though we have learned to pinpoint exact locations on the globe and transport ourselves there in a matter of hours, we may still admire a tiny bat who achieves a comparable feat unassisted by maps and machines. Bats can find each other across vast distances and dense crowds. Contrary to myth, no bats are blind, but these caves are very dim by day and pitch dark at night. Imagine finding your partner at a cocktail party with a million guests where everyone, including you, is blindfolded.

Earlier studies had shown that Mexican free-tails know their pup's individual scent, and learn the spatial geometry of the cavern where they gave birth. And just as salmon famously do, they also frequently return to their natal cave to have their babies. Observations of mothers and pups tagged with a dab of infrared reflective paint (visible in the dark to a special camera but not to the bats) show that the pair do indeed reunite, and that the mother usually lands within a few feet of her pup.[2] Pinkies don't migrate very far in the daily jostle of tightly packed bats, so the mother narrows her search task to a few thousand bats by landing in the vicinity of where she last left her baby.

My project was to investigate how vocal recognition helps mothers reunite with their pups. For three summers I captured nursing pairs, recorded their calls both inside and outside of the caves, conducted playback experiments, and released the bats where I had found them. I devised a circular arena with an entrance tube leading to the midline demarcating the left and right halves, and a pair of speakers at opposite sides issuing the plaintive (and to human ears mostly inaudible) cries of two bat pups. After trials with thirty different bats, it was clear that a mother spent significantly more time near the speaker broadcasting her own pup's calls than near the other speaker.[3]

I spent many hours analyzing my recordings of their calls. The bats' calls are partially audible to the human ear, but our brains process sounds at much slower speed and lower frequencies. By tape-recording their calls at high speed and playing them back much more slowly, a whisper of faint blips and squeaks resolved into a constellation of unique cries. Perceptually, it was as if my hearing had become the mother bat's, and I could now distinguish the insistent bellows of perhaps a dozen pups who happened to be within range of the directional microphone. Each baby bat's call-signature was markedly distinctive. One's cry might start low and spiral upward, wavering twice before dipping again at the end. Another pup's call might begin with a fairly constant tone, then drop dramatically before rising at the end. The frequency-time sonograms in Figure 1 illustrate visually the distinctive character of these pups' calls.

Each call lasts about a tenth of a second. Real calls are tonally richer because they include several harmonics. When the calls are represented on a graph, each bat's calls can be seen to be consistent and as unique as a signature. These are the hallmarks of a finely tuned individual recognition system honed by millions of years of requiring that individuals find each other in dark, crowded conditions.

The reunions between mothers and pups are only one, albeit vital, facet of the lives of Mexican free-tailed bats. They have long, narrow wings built for speed. Many times I stood at the cavern entrance watching them make their evening exodus. It is a spectacular sight. They begin making circuits near the cave mouth a half hour before they emerge. Their numbers swell and just when it seems there's nowhere else to fly but out, the bat cyclone suddenly pours forth, like tea from a spout. It can take two hours for all of them to make their exit.

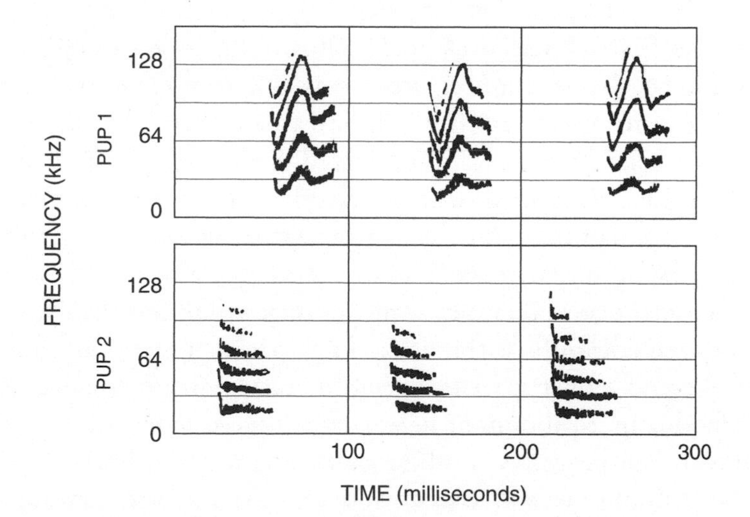

Figure 1 Sonogram of two bat pups' calls

While watching the multitudes in flight, it is easy to forget that each one is a breathing individual, unique from all the others. I handled about a thousand free-tailed bats, and grew to love their wrinkly, pouting muzzles and alert eyes tucked below large, heavy ears. Their short, fine fur felt as soft as down. Scent glands on their heads give off a pungent perfume, and left an oily residue on my fingers after handling a few dozen.

For the first mile or so they stay tightly bunched and the column snakes away into the sky, undulating like a dragon in a Chinese New Year parade. The bats' cohesiveness is thought to be a defense against predators; one is less likely to be singled out in a crowd, and aggregations are thought to confuse predators. But the strategy is not foolproof. At James River Cave, where I was based in 1989, a red-tailed hawk made regular swoops into the bat column. She was a skilled and experienced bat-snatcher, sometimes nabbing one bat in each foot on a single pass. With each catch, she would make a beeline over the cliff from where she'd appeared. She surely had an active nest there, for a few minutes later she would reappear with empty talons to launch a new ambush. Each individual capture spelled almost certain death for two bats—the mother snatched by the hawk and a little pup somewhere in the cave who had just lost his or her lifeline.

In three years of watching the evening exoduses of the bats, I had never once watched them return. On my final day in the Texas outback, I rose before dawn and trudged out to the cave entrance. It was quiet at first. Then, as the eastern sky began to tint with first light, I heard the first whizzes and buzzes of tiny forms zooming into the cave entrance. A few minutes later I was able to point my binoculars skyward and see them arriving. Bats streamed across the sky from all directions, sparsely at first. They folded their wings and practically fell out of space. As they neared the cavern's mouth, they opened their wings slightly to brake and steer, causing a ripping sound as the air buffeted their wing membranes. These spitfires, weighing just 14 grams, had flown as much as 60 miles away, yet they still found their way back to a pinpoint location in the dark. They perform the same feat over greater distances on their late spring migrations from Mexico, where insect food is more abundant in winter. It isn't known quite how they do this, but I suspect they combine geomagnetic and celestial compassing, recognition of topographical landmarks and, when they approach the destination, their sense of smell.

Mexican free-tailed bats emerge from a nursery cavern in southern Texas. (Photo by the author.)

From Knowledge to Change

The evolution of sentience—the capacity for pain and pleasure—was a crucial turning point in biological history, affecting all animals. Before sentience, living organisms had no moral consequence, for two reasons. First, an organism without feelings cannot suffer. Second, eons had yet to pass before there would be any highly evolved minds to reflect on moral matters such as the rightness or wrongness of an action. But for a sentient creature, things can be perceived to be going well or poorly. My experiences studying and observing animals "up close and personal" reinforced what I had long held to be self-evident: animals experience pain, pleasure, and emotions, and their lives have meaning beyond any utilitarian value that humans may place on them. Morality didn't originate with humans (see Chapter 8), but acute moral awareness is one of humankind's greatest achievements. It is also one of our heaviest burdens.

Getting to know animals has also made me more aware of the immense chasm between the way we ought to treat animals and how we actually do treat them. Our relationship to animals has been based mostly on a "might makes right" ethic. According to the Food and

Agriculture Organization of the United Nations (FAO), over 50 billion land animals worldwide were killed for food in 2005.[4] The number of individual fishes killed by humans may be higher still. Close to 100 million mice, rats, rabbits, monkeys, cats, dogs, and birds are consumed yearly in American laboratories.[5] Between 40 and 70 million mourning doves are shot by American hunters yearly.[6] The Humane Society of the United States (HSUS) estimates that over 50 million animals are killed for their fur each year around the world.

Advances in our knowledge of animal sentience are compelling humans to reconsider our prejudices toward animals. One of humankind's greatest strengths is the speed with which we can undertake profound cultural change. Our species' significant advances over institutionalized racism and sexism represent two of our most admirable social achievements (albeit with work still to be done). That these developments have occurred during the past two centuries—an eye blink in the one- or two-million-year existence of our species—illustrates how quickly profound social change can happen when an injustice is laid open for scrutiny.

Ethical concern for animals as a serious moral issue has roots in the eighteenth and nineteenth centuries, but it didn't gain significant momentum until the latter part of the twentieth century. Today, "animal rights" is a fairly universally recognized term in Western cultures. A Google search yields over 66 million results.

The problem in our relationship with animals is that our treatment of them hasn't evolved to keep up with our knowledge. While we have banned bearbaiting and passed some animal welfare laws, animals' position relative to a sphere of moral consequence remains unchanged: they are outside it. As long as we provide a reason that our harming them is "necessary"—for example, to eat them, to make garments from them, to use them in tests for human safety, and so on—then we may do so, even though we do not accept using humans in these ways.

Let me say at the outset that I will not try to convince you that animals are merely other manifestations of humans. Each species is unique, there are marked distinctions across species, and some of the strongest distinctions involve humans. But, as Charles Darwin famously said, these are differences of degree, not kind. And how those differences translate into how humans treat animals is well worth our careful consideration.

CHAPTER TWO

tuning in: animal sensitivity

All animals are equal
but some animals are more equal than others
—George Orwell, *Animal Farm* (1945)

The world is a dynamic place. Seasons and weather patterns change; animals, both friend and foe, move about the landscape; plants flower or bear fruit while others are dormant. Animal life is demanding, and their environments require them to be prepared to search for food, avoid becoming food, find mates, seek shelter, migrate, and maintain contact with companions. Given that they have evolved in a diversity of environments—murky, dark, bright, noisy, crowded—they've developed some pretty spectacular ways to orient, survive, and thrive. In the course of a few hundred million years, animals' sensory systems have evolved to function with astonishing efficiency. Humbling as it may be, for all our vaunted brain power, humans emerge as nothing special in the sensory sweepstakes. Our senses of vision, hearing, smell, taste, and touch are middling, at best.

Sentience Expanded

How does a human's sentience—our capacity to feel pain and pleasure—stack up against a nonhuman's? An important problem with questions like this is that we cannot know for certain, because another's feelings, whether simple or complex, are private. We can, however, divine a great deal from the anatomy, physiology, and behavior of other animals, and by using our own experiences as a guide. Insofar as we are more intelligent than, say, a sea lion or a bat, we may be capable of richer experiences and feelings in the mental-emotional domain. I can anticipate getting together with friends, journeying to a new country, or delighting in a clever joke, whereas the sea lion, presumably, is not privy to at least some of these sorts of mental pleasures. But to the same degree, the sea lion is perhaps comparatively free from the mental anguish my own rational mind is capable of producing. In his best-selling book *The Power of Now*, the spiritual teacher Eckhart Tolle seeks to guide us out of the angst and unhappiness we bring on ourselves by our preoccupations with unalterable past and unpredictable future events.[1] On this reasoning, there is no clear basis for the assumption that a more intelligent life is inherently better (or worse) lived. Quality of life does not align smoothly with intelligence.

Furthermore, a less intelligent animal may experience life no less richly than a human, in the sensory realms. This is not a new idea. The British clergyman Humphry Primatt (1735–1776), an early writer on animal welfare, knew the tenuousness of linking suffering to intellect: "Superiority of rank or station exempts no creature from the sensibility of pain, nor does inferiority render the feelings thereof the less exquisite."[2] More recently, the biologist John Webster, author of *Animal Welfare: Limping towards Eden,* has expressed a similar sentiment, if somewhat more bluntly: "People have assumed intelligence is linked to the ability to suffer, and that because animals have smaller brains they suffer less than humans. That is a pathetic piece of logic."[3]

Because animals evolve greater sensitivities in realms relevant to their survival, and because different niches present diverse sensory challenges to organisms, it follows that humans have not cornered the evolutionary market on sensory perceptions. Elephants communicate in infrasound, bats in ultrasound, and some fishes with electricity. Many organisms far exceed us in olfactory and other chemical sensitivity. Others have more advanced organs for detecting subtle changes in water or air movements. In the emotional domain, it is far from clear that a monkey's or a rabbit's fear is felt less acutely than our own fear, or that feelings of affection, and subsequent grief at their loss, are duller between two parrots who mate for life than such feelings between two humans.

As far as is known from physiological studies, the perception of noxious stimuli and their conduction to parts of the brain that register pain are fairly identical processes among the different mammals that have been so examined.[4] The benefit of measuring pain perception at the brain level is that it is not vulnerable to confounding factors such as stoicism, which may make an animal appear to be less in pain that she or he actually is. Age, genetics, sex, and prior experiences are other factors that may influence the perception of pain. In a paper published in 2008, University of Massachusetts veterinarian Jerald Silverman concluded that "with regard to experiencing pain, there are no unequivocally 'higher' or 'lower' sentient species among the mammals."[5]

Because many animals have more acute senses then we do, they may feel certain things more intensely than we do. What proof have we that a needle prick is less painful to a mouse than to a man? Recognizing a painful sensation and trying to escape from it should not be any less compelling an evolutionary imperative for a rodent than for a primate. The propagation of mouse genes—which pain evolved to assist by helping mice avoid situations that threaten to destroy them and their genes—is no less worthy a project for a morally indifferent nature than is the propagation of human genes. In some situations, it is possible that a human's knowing the reasons for pain—such as a necessary medical procedure—may lessen (or intensify) the experience of the pain.

British ethologist Donald Broom believes fishes may in some cases suffer more than we do, for they may lack ways that we have for dealing with pain. For instance, humans can be told (or we can tell ourselves) that a pain will not last for long, whereas fishes presumably are unable to do so.[6] For American ethologist Marc Bekoff, suffering may be greater in an animal with no rich cognitive life with which to remember past events or anticipate the future.[7] In *The Unheeded Cry: Animal Consciousness, Animal Pain, and Science,* American bioethicist Bernard Rollin suggests that animals with a reduced concept of time may not look forward to or anticipate the cessation of pain. "If they are in pain, their whole universe is pain; there is no horizon; they are their pain."[8]

It is well documented that the adrenal response to stress in humans is not nearly as dramatic as in animals. One possible explanation for this is that humans are very adept at finding ways to cope with stress. Our various psychological defenses and coping devices "damp down" the stress response.[9] When I reviewed published articles on animals' responses to unpleasant or painful routines in laboratory settings, I learned that their stress responses are pronounced and lasting. In response to being stuck with a needle, having blood drawn, or being force-fed, rats, mice, rabbits, monkeys, and various birds all show dramatic increases in typical stress markers. Blood levels of the "stress hormone" corticosterone soar to as much as five-times normal, and it can take up to 90 minutes before they return to baseline levels. Heart rates and blood pressure also rise.[10] There are also so-called witnessing effects: Animals who are in the same room as other animals being mistreated or killed also show stress reactions.[11]

Sensitivity to pain has been better studied in farmed animals than most other animals. Physiological and behavioral responses to such routine practices as castration, hot-iron branding, tail removal, horn-bud cauterizing, and beak-searing—all of which are performed without anesthesia—indicate that pain is intense and lasting.[12] Though "prey" animals such as sheep, cattle, goats, and pigs may have evolved stoicism to avoid being singled out as weak or vulnerable by a lurking predator, changes in body posture and movements (e.g., shaking, twitching, alternating lifting of hind legs), and raised blood stress hormone levels indicate that their pain is real.

Know Thine Umwelt

It is a fantasy of mine to have the power to become another animal in order to experience life from their perspective. I generally choose to retain my own mind but within an animal's body, and I tend to favor birds and fishes as the subjects of my metamorphoses. The German ethologist Jakob von Uexküll also thought about what it would be like to be another creature. Around 1905, he coined the term "umwelt" to refer to an animal's sensory world. The idea is that variations in brains, sensory equipment, and lifestyles of different kinds of animals likely result in their having different mental and perceptual experiences. Dogs, for example, see mainly in black and white, but their acute sense of smell allows them to discern a kaleidoscope of information. Just watch dogs on their walks: they spend a lot of time with their nose against the ground, sniffing up clues as to who or what has been there before.

Once, when I was studying bats as a graduate student, I was standing in an indoor flight cage with several horseshoe bats hanging from the ceiling in the opposite corner. I was about ten feet away from them. With my

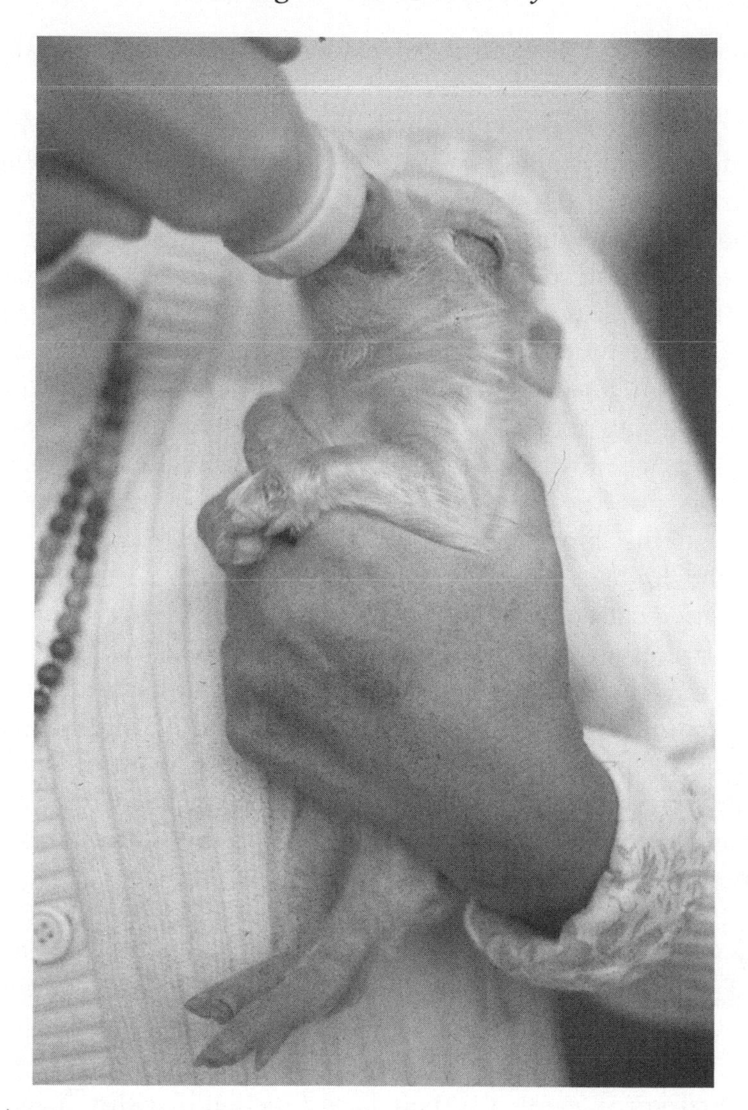

A differently intelligent animal may experience life just as intensely as does a human. (Photo by the author.)

hand at my side I rubbed a finger softly against my thumb. I couldn't hear the sound produced, but the bats' oversized ears immediately snapped to attention, funneling in my direction and quivering rapidly. When I stopped rubbing, their ears relaxed, but they locked on again the moment I resumed. The subtle stimulus I was producing contains ultrasound, and though it was surely faint, these bats, schooled for hundreds of millennia in the fine art of ultrasound detection, could hear it.

Echolocation is a fantastic adaptation, but it has its drawbacks. Ultrasound dissipates very fast in air, so most bats shout very loudly to detect their own echoes. How, then, does a bat's ear detect a faint echo immediately after being bombarded with a high-intensity, potentially damaging pulse of sound that may exceed 140 decibels, beyond the pain threshold of human hearing? (Fortunately for us, their call frequencies fall outside our umwelt—they are too high for us to hear.) The solution is as strange as it is simple: bats disable their hearing during the call phase and reactivate it during echo detection. Among the fastest rates of muscle contraction are found in the ears of bats, where muscles may twitch 120 times per second to temporarily deafen the bat.[13]

Echolocation also serves another communication function, one that the echolocator (the bat) doesn't even intend. These vocalizations may contain useful information for other animals who can hear them. Sure enough, there are many examples of others eavesdropping on the echolocation calls of bats.[14] For instance, red bats in southern Ontario pursue larger insect prey, such as moths. Our playback experiments showed that these bats were attracted to the rapid bursts of echolocation ("feeding buzzes") of other red bats made during the final pursuit on a moth. This, combined with our observations that two bats occasionally attacked the same moth at once, supports the idea that a red bat may eavesdrop on another's feeding buzzes in an attempt to pirate the moth for him or herself.[15]

If you think the moths are resigned to being victims, think again. Many moth species have evolved their own bat-detection system. If a pair of eardrums on their thorax picks up approaching echolocation, the moth will take evasive action. Some species dive-bomb for the ground, and some adopt an erratic, spiraling flight, while others produce their own ultrasonic signal that is thought to function by either "jamming" the bat's sonar or by startling the unsuspecting bat. By shaking one's keys at moths, one can observe the reaction of species tuned into bat echolocation; they dive-bomb or (if not flying) shudder their wings in response to the ultrasounds produced by the colliding keys.

There are innumerable other dynamic acoustic interactions in the vast world of bats, none of which humans are normally privy to because the action falls outside our umwelt.

When we study them closely, we often find animals capable of feats that we would not have imagined possible, mainly because we lack those capacities ourselves. Migrating birds can actually "see" the earth's magnetic field, which is generated by the movements of molten metals beneath the earth's crust. Studies of the migratory garden warbler show specialized neurons in the eye that are sensitive to magnetic direction, allowing the birds to perceive the magnetic field as a visual pattern.[16] In August 2008, five European scientists announced that cattle and deer across the globe, whether grazing or resting, align their body axes in roughly a north–south direction.[17] The reason is not clear, but it boggles the mind that no one—herdsman, rancher, or hunter—had noticed this before. What else are we failing to notice?

As we'll see later when we delve into animals' social interactions, animals are acutely aware of others in their midst. Domesticated dogs more readily approach the winner of a playful interaction between another dog and a human than they do the winner of a competitive interaction (i.e., lacking play signals). This suggests that playful individuals are more desirable as social partners than are competitors. In experiments, dogs unable to observe an interaction between a human and another dog still respond differently to winners and losers, indicating their ability to "read" the interaction and its outcome from the demeanor of the participants.[18]

Time Perception

Have you ever wondered if other animals experience time at a different rate than us? The perception of time by animals with different life spans and activity rates was another part of the umwelt concept conceived by von Uexküll. Popular folklore assigns seven dog years to every one of ours, but this seems likely to be more a product of dogs' shorter life span than their actual *experience* of time. Yet, the speed of some animal responses reveals a finer perception of time than we can achieve. Knifefish communicate with electrical discharges of up to a thousand pulses per second. A nightingale sings each note of his elaborate song in just one-tenth of a second; humans can only appreciate its complex nuances if they record it at high speed and slow it down.[19] Mated pairs of Amazon splash tetras make synchronized leaps from the water to deposit and fertilize their eggs on overhanging leaves. So closely timed are these leaps that the two fish appear as one.[20]

I have timed the quick movements of a brown creeper shimmying up the trunks of trees and vines, turning her head and probing the crevices with her beak. Each movement occurred at a rate of approximately three to four per second. If this tiny bird is using her brain consciously with each movement—as seems likely, given that she is foraging and must detect and respond to specific cues in her surroundings to do so successfully, all the while remaining vigilant for the occasional lurking hawk—then her perceptual speed must function faster than ours.

Bats have been shown in experiments to be able to make time discriminations in their perception of echoes down to 10 billionths of a second.[21] Some people are incredulous at this, but we may expect animals to be exceedingly good at perceiving stimuli that are highly important to their survival. Being able to make precise time discriminations in three-dimensional space in the dark can mean the difference between a hit and a miss for a little brown bat (*Myotis lucifugus*) pursuing a midge.

Flocking birds and schooling fish are renowned for the coordination of their movements. To our eyes at least, it appears that the entire flock or school changes direction at one instant, as if there were some internal knowledge of the decision-making of all others. Some naturalists had ascribed this coordinated behavior to a form of telepathy, but analysis of slow-motion filmed sequences shows that these creatures' sensory systems are operating on a much finer time scale, so that it only appears to us that they all change direction instantaneously when in fact there are minuscule delays.

Many birds sing antiphonally. In some species of wren, up to eight birds intercalate individual calls into a unified, seamless whole that has stunning acoustic beauty and coordination. Plain-tailed wrens (*Thryothorus euophrys*) of South America sing in synchronized choruses, each bird using a repertoire of about 20 phrases, each carefully matched to the other birds' calls. Songs can last two minutes, during which individuals drop in and out.[22] This speaks to the birds' remarkable temporal and pitch perceptions. Human encroachment may be forcing the wrens and other birds to ramp up their signals. Dutch researchers recorded the calls of great tits in ten major European cities, including London, Paris, Amsterdam, and Prague, and found that city-dwelling birds sing shorter, faster, and higher songs than the slower melodies of their country cousins.[23] It's thought to be an adaptation to counter background noise and make it easier to find a mate amid the urban din. Umwelts evolve.

Beyond Our Senses

To get by in their ecological niches, many animals have evolved perceptual abilities that exceed our own. Wildebeests have the uncanny ability to detect the presence of rainfall thirty miles away. Using this ability, wildebeests move to greener pastures, rather than waiting for the rain to come to them by chance. How they do this is not certain, but researchers studying them in Kenya have speculated that they may use a combination of sight, smell, and sound.[24]

Because seed predators hinder a tree's reproductive output, many trees have evolved a swamp-and-starve strategy, producing relatively little seed in most years, then a sudden glut before the seed-eaters can recover. Squirrels, however, are somehow able to predict a "mast" year when a spruce tree produces a bumper crop of cones. In such years, mother squirrels interrupt the weaning of their first litter of pups and conceive a second one. They do this ahead of the impending food glut, and it isn't known quite how they anticipate the bumper crop; perhaps they are able to see a difference in the buds that will form cones, or perhaps they can detect a chemical change.[25] Either way it allows squirrels to hear the dinner bell before it has even been rung. Mangabey monkeys have been shown to rely on their memory of recent patterns of temperature and solar radiation to decide whether or not to travel to a particular patch of fruit.[26] Biologists working on Panama's Barro Colorado Island claim that if you want to be sure a storm is nigh, just listen for the howls of howler monkeys. They never get it wrong.[27]

Estuarine crocodiles have superb long-distance homing abilities. This was not known until scientists decided to affix tracking devices to three adults caught near popular Australian beaches and rivers before shipping them to more remote areas. For instance, when one large crocodile was flown by helicopter to the east coast 400 km (250 miles) away, he was back home within three weeks.[28] One scientist couldn't resist likening them to boomerangs, except that boomerangs don't think or act on their own.

While there is probably a physical explanation for these phenomena, they illustrate certain types of perception that far exceed the capabilities of humans. Perhaps these animals have sophisticated ways of predicting future patterns based on past events combined with a rich mental representation of their physical environments.

Because we don't think or see like them, and because we are not intimately versed in another species' postures, vocalizations, smells, and personalities, we miss a lot of what's going on. Fortunately, technological advances are expanding our observational niche. It was only when the play of babbler birds was studied on slow-motion video that subtle exchanges between individuals were noticed: eye contact and special postures during play, for example.[29] Similarly, Marc Bekoff only discovered the role of eye contact and stances in the play of dogs when he examined video frame by frame.[30]

Research from the University of Sheffield and the Massachusetts Institute of Technology has shown strong parallels between rats' use of whiskers and humans' use of fingertips to explore their surroundings. High-speed video recordings of the movements of the animals' whiskers and their associated muscles reveal that rats adjust their whisker movements, "whisking" them back and forth many times each second, using information from each contact to decide how best to position the whiskers for the next one. Smooth surfaces are explored with periodic waves of motion, while rough surfaces are treated with large, irregular, high-speed brushes. These movements are actively controlled by the rats, just as we guide the movements of our fingertips as we explore the feel of shapes and textures.[31]

Walruses also have supremely tactile whiskers. Recent studies show that walruses and manatees control their thick vibrissae by a network of muscles arranged like the struts of the Eiffel Tower. Each whisker can be telescoped out and moved in a coordinated fashion, allowing the animal to detect the size, shape, texture, and taste of small objects hidden beneath mud and sand.[32] Colleen Reichmuth of the University of California at Santa Cruz reports that "if you drop a little piece of fish on the whiskers away from the mouth they can walk it along the whiskers, across the muzzle and into the mouth."[33]

Mouse Fidelity

Cute and popular in animated films, reviled as "pests" and abused as laboratory subjects, house mice pay royally for cohabiting with *Homo sapiens*. Yet cohabit they do. The temptations are just too great, and they have enough guile to thrive. Rodents, in particular mice and rats, are also commonly demeaned as "lower mammals," even though molecular studies suggest that rodents (and rabbits) are more closely related to monkeys than are dogs.[34]

Most humans around the world have had contact with mice. Extraordinarily resourceful and successful at living commensally with us, the house mouse in particular has made a good living by entering human habitations and living off the tailings left by their lumbering, distant mammalian cousins. Small and with a straightforward "design," the mouse ranks, on first glance, as a cheap compact sedan among some of the Cadillacs and Bentleys of the mammalian class. Yet, on closer inspection, the mouse is an astonishing marvel. House mice are acutely attuned to their surroundings. They are highly sensitive to sound, smell, taste, and touch. Their vision, while relatively poor in brighter light, is keenly adapted to the low-light conditions they encounter in their nocturnal wild ways.[35]

Their sensory worlds are quite different from ours. If we left a trail of urine wherever we went, we'd not be very popular, but for a mouse, it's a normal social courtesy. A mouse's urine is like a signature. It contains chemical information that communicates an individual's sex and social status.[36] Mouse pee also allows other mice to discern genetic relatedness, a process which may have evolved to avoid inbreeding.[37] There is even evidence that female mice can discriminate the degree of parasite infestation in males based on the smell of their urine, and that this may in turn influence females' mating proclivities.[38] It is not clear whether mice are conscious of the discriminations they make. At the very least, though, these abilities indicate that a mouse's sensory system operates with fidelity comparable to a human's.

In 1967, a six-year study of captive white-footed mice was published by UCLA biologist Lee Kavanau. Interested in the activity patterns of these small mammals, Kavanau built them an indoor habitat with an extensive system of tunnels and a large, automated running wheel. The mice were extraordinarily active. Being nocturnal beasts, they spent much of the night eating, drinking, and exploring. On average, the running wheel went through 33,000 revolutions per night. Their human observer surmised that one of the rewards of wheel running is its requirement for split-second timing and coordination of movements and quick reflex actions.[39] It also enabled the mice to act like their counterparts in the wild, who typically eat about two hundred small meals nightly, returning to some twenty to thirty food sites.

Perhaps most remarkable was the spatial learning and memory demonstrated by the little rodents. Their complex habitat included 427 meters of simulated burrows constructed of connected tubes. This vertical maze system had 1,205 ninety degree turns and 445 blind alleys, and the shortest one-way path from end to end was 96 meters. It took individual mice just two to three days to learn to traverse this complex system—both forward and reverse—without any associated reward or prior deprivation (methods often used to motivate animals to do what researchers what them to do). Based on the speed with which the mice learned elaborate sequences of lever presses coupled to programmed sound cues, Kavanau concluded that mice learn complex experimental regimes with a facility comparable with nonhuman primates. Kavanau's data show that animals genetically pass on to their offspring the tendencies to explore, contrary to the traditional view that hunger and thirst motivate such locomotor movements.

These animals are naturally predisposed to vary their behavior patterns. For instance, on a typical night, a given mouse obtained fifty pellets using the correct sequence of maneuvers required to get each pellet. The mice also failed an average of twenty-three times. These were not twenty-three errors, *per se,* but rather a deliberate natural expression of behavioral variability—a crucially adaptive penchant for living in a complex, changeable world. When Kavanau mimicked such changeability by resetting the required sequence to obtain a pellet, the mice quickly discovered the new sequence. Had they been doggedly fixed to the prior one—had they lacked resilient intelligence—they might have gone hungry.

Kavanau observed another intriguing pattern of behavior. The mice showed an almost obstinate insistence on exercising control over their environment. Provided with switches to control their own light levels, the mice selected dim light for their active periods, and very dim light for their inactive periods. These are roughly the same illumination patterns they would have in nature, where they hide away in their dark burrows and nests by day. Kavanau noticed that the mice tended to resist "with astonishing vigor" being forced to do something. Whenever he handled one and returned it to its nest, the animal would immediately leave the nest. And no matter how motivated they were to run, the mice would switch off the running wheel if someone switched it on, and vice versa. An act was rewarding when done volitionally, but not so when initiated by force. The conclusion: mice value self-actualization. This shouldn't be too surprising. Animals in the wild are seldom forced to endure conditions from which they can not escape or whose severity they cannot reduce. Alas, in a typical laboratory, that is routinely the case.

I have only mentioned two species of "mice" here: house mice and white-footed mice. Worldwide, there are several hundred species of small, mouselike rodents. Voles alone comprise about 124 different kinds. We can only wonder what diversity of behavior and perceptiveness these animals exhibit that remains unknown to us.

Experiencing Instinct

René Descartes argued that animals were ruled by an inflexible instinct that cannot aid them in different circumstances. To this day we have a common tendency to reduce animals' actions to mere instinct.

There is nothing "mere" about instinct. Babies are instinctively drawn to movement and faces. Men are instinctively drawn to women with symmetrical faces and a big waist-to-hip ratio. Women are instinctively drawn to men with a major histocompatibility complex (a dense area of our genetic makeup that plays an important role in immunity) vastly different from their own. Instinct keeps us alive (swerving to avoid a tree), safe (avoiding precarious heights) and intact (removing your hand from a hot element). In a complex world, it would be too time-consuming (not to mention dangerous) if minds had to consciously, rationally, process all the information they are confronted with.[40]

But what is crucial to understand about instinct is that it does not (as is so often assumed) preclude conscious experience. When you jump at a sudden noise, or you blink when something suddenly flashes before your face, you still experience and remember these events, even though your body responded involuntarily. All of the basic feelings we experience are fundamentally instinctual. We don't intellectually learn that itchy feelings are unpleasant or that the sweetness of a blueberry is attractive.[41] But only a clever robot would deny that a mosquito bite or a fruit smoothie is consciously experienced.

An example of a learned behavior that might initially appear instinctive is the way grasshopper mice avoid being sprayed with the hot chemical defense of the bombardier beetle. Grasshopper mice inhabit southwestern American deserts and prey on a variety of invertebrates. A young, naive mouse is in for a nasty surprise when he takes on a bombardier, whose rear end delivers a caustic blast of chemicals. Coping with these beetles is not instinctive; it must be learned. Grasshopper mice quickly learn to associate the appearance of bombardier beetles with their noxious defense, and they soon deploy an effective counterattack. Seasoned mice attack the beetle's head and quickly thrust its abdomen downward where it discharges harmlessly into the sand. It is clear that this behavior could only have developed in a conscious, aware mouse rather than in an automaton governed only by instinct.

I remember encountering a killdeer acting strangely while I wandered a riverside park in southern Ontario many years ago. When I first saw the bird trotting along haltingly with one wing extended awkwardly from his tilted body, my automatic reaction was that he must be injured, and it was only when I remembered having seen a documentary film of the behavior that I realized what was going on. Killdeers and other plover species are renowned for this ploy of feigning injury, which is performed by both sexes and functions to lure would-be predators away from their vulnerable ground nests.

The killdeer's broken-wing act was once written off as an example of mindless instinct, but it has since been subjected to detailed study in the field. Led by biologist Carolyn Ristau, a team of investigators played the role of "predators." Here is what they found. The threatening interloper is monitored closely: killdeers perform the behavior in open areas where they are more likely to be seen, and they deliberately situate themselves in the line of a predator's sight. They often look back at the predator, and if their actions fail to elicit a response, they fly closer and display more vigorously. In a series of experiments designed to test whether killdeers can tell the difference between more threatening and less threatening animals, a pair of human volunteers dressed in distinctive clothing approached the bird's nest in succession. One volunteer stared menacingly at the nest while walking past, whereas the other ignored the nest. Later, when each volunteer was presented to the bird in random order, the parent bird responded more strongly in 25 of 31 such trials (leaving the nest) to the human who had behaved more threateningly. Other observations have shown that killdeer recognize cattle as having benign intent, but that they are nevertheless possible threats because they could accidentally step on a nest. The plovers respond adaptively to this threat by giving warning cries rather than by displaying a pointless pretense of injury. Killdeers act as any conscious, feeling creature would in the face of danger to their young. They take risks, and they adapt flexibly to the particular challenges of the situation.

The dam-building of beavers is another instinctive behavior that shows considerable flexibility. Beavers inherit instinctive knowledge of basic techniques of dam and lodge building, but they won't build if they don't need to; they will adopt suitable buildings on shore, and use human-built dams.[42] Just as no two predator threats are the same to a killdeer, so too are no two dam-building projects identical. Beavers use a variety of materials according to availability and the particular building situation. They may modify human artifacts, for example adding mud, sticks and even stones to a concrete dam, thereby raising the water level from that intended by humans. They also build lodges and nesting chambers inside caves or human buildings such as unoccupied or occupied buildings near streams, or unused mills.[43] Donald Griffin called these modifications of behavior "sensible use of available resources rather than the unfolding of rigid, stereotyped genetic programs."[44]

In its pursuit of rigor and economy of explanation, favoring the simplest of plausible theories, science places the burden of proof on those who would ascribe thoughts and feelings to animals rather than on those who would deny animals these attributes. With the overwhelming accumulation of evidence showing that animals think and act flexibly in response to changes in their surroundings, it's time to shelve the old dogma. The question is no longer Do animals think? but What do animals think? Moreover, we should humbly allow the likelihood that animals have more going on in their minds than our limited vantage points allow us to appreciate. As the psychologist Nicholas Humphrey has said, "if a rat's knowledge of the behaviour of other rats were to be limited to everything which [animal] behaviourists have discovered about rats to date, the rat would show so little understanding of its fellows that it would bungle disastrously every social interaction it engaged in."[45]

CHAPTER THREE

getting it: intelligence

If men had wings and bore black feathers,
few of them would be clever enough to be crows.
—Rev. Henry Ward Beecher (1813–1887)[1]

We are an "intellicentric" species. We applaud geniuses, and we have a rich vocabulary to disparage idiots, imbeciles, morons, twerps, nitwits, boneheads, dunces, ignoramuses, and simpletons, to name just a few. Intelligence is a complex dimension that doesn't just occupy a simple scale of low to high. You probably know someone who excels at solving physics problems but is a poor wordsmith; another friend may lack skill in either of these realms but be gifted with eye-hand coordination. I have long-term friends whose twenty-five-year-old daughter suffers from a serious neurological condition; she is academically slow, but a gifted and creative musician and her first CD of piano compositions is among our favorites in my home.

Comparing the intelligence of different animals is like comparing their ability to move. Do fish move better than horses? asks Herbert Roitblat of the University of Hawaii. Animals are as intelligent as they need to be. If a particular mental ability—such as learning to recognize other individuals, or to identify predators—is important to survival and reproduction, then it will be favored evolutionarily. But nature doesn't waste energy building brains just because it can. All else being equal, an organism with a smaller brain should have a survival advantage over one with a larger brain, because the "brainier" one must consume more energy to sustain its gray matter.

Measuring Up

We have a patronizing tendency to measure other animals' intelligence against our own. We subject them to a sort of IQ test. Because these tests are devised by humans, they naturally contain human bias. Counting oranges in a box may reveal something of the intelligence of a chimpanzee or a parrot, but it sheds little light on the animal's intelligence. Counting fruit is not very useful to them. Recognizing edible fruits, knowing where they are in a forest and when they are due to become ripe—these are useful bits of information to a chimp and a parrot.

Animals tend to excel at things important to their survival and success. Merriam's and Great Basin kangaroo rats both stockpile food, but whereas the seed-eating Merriam's stores them in scattered locations, the leaf-eating Great Basin species tends to hoard its supply more centrally. Merriam's tend to have to venture farther afield to find their food, so storing them in smaller caches near where they are found is more efficient and less vulnerable to pilfering by other seed-eating scroungers. Now, guess which species performs better at relocating four predetermined cache sites after a delay of 24 hours? If you guessed Merriam's, then give yourself an extra sunflower seed.[2]

Just for the record, humans are not always smarter than nonhumans. The starkest demonstration of this was discovered at Kyoto University, Japan, where a community of captive chimpanzees participated in various studies of cognition. Tetsuro Matsuzawa, who directs the program, presented the chimps with touch-sensitive computer monitors, and they soon learned to obtain small food treats by performing tasks on the monitor. One such task is to recall in sequence the numbers 1 to 9 scattered randomly on a computer screen for just one second before being replaced by white squares. Ayumu, a five-year-old chimp, excels at this. He casually but quickly points to all the white squares in sequence.[3] Humans barely pass the test with just four or five numbers. When the British memory champion Ben Pridmore—who can remember the order of a shuffled deck of cards in 30 seconds—competed head-to-head against Ayumu, the chimp performed three times better. When the numbers flashed for just a fifth of a second, Ayumu correctly recalled all nine digits 90 percent of the time, compared to 33 percent for Pridmore. While Ayumu is the best among his chimp peers, the average chimp scores twice as well as the average human on this short-term memory task. It appears our hairy ape cousins have a photographic memory.[4]

Matsuzawa believes that chimpanzees' complex social networks explain their aptitude with this task; in the wild, it is very important for individuals to keep track of the locations of other individuals in their group. Wild chimpanzees also demonstrate a "botanist's memory" for up to two hundred plant species; they know what each plant is used for, its location in the forest, and the approximate timing of fruit availability.[5] American anthropologist Jill Pruetz notes that "observing that other species can outperform us on tasks that we assume we excel at is a bit humbling.... We should incorporate this knowledge into a mindset that acknowledges that chimpanzees, and probably other species, share aspects of what we think of as uniquely human intelligence."[6]

Thinkers

A captive dolphin named Kelly who is kept at the Institute for Marine Mammal Studies in Gulfport, Mississippi, has developed behaviors that involve reasoning, tool use, planning, and deception all rolled into one. The human trainers there decided to recruit the dolphins to help clean up litter that makes its way into their pools. This is simple enough, for they soon learn to trade litter for fish. But Kelly soon learned to manipulate this arrangement to her own ends. When a piece of paper contaminates the pool, Kelly carries it to the bottom and hides it in a rocky crevice. Next time a human trainer walks by, she tears off a piece of paper and trades it for a fish! This involves delaying gratification (by resisting temptation to hand over the paper when she first gets it) and the realization that a small piece of paper is of equal value to a whole sheet.[7]

An animal with this sort of smarts isn't likely to stop there. Before long, Kelly was delaying gratification again—this time stashing hunks of fish to lure gulls to the water surface. Then, with the stealth of a pickpocket, she would snag the birds' feet in her jaws and trade their freedom for another fish hastily delivered by one of her trainers. She taught her calf to do the same trick, and gull-baiting became a hot new game among the dolphins.

This is not the first time that a dolphin has turned the tables on the teacher-pupil relationship. Science writer Eugene Linden describes a doctoral student's experiments with a captive dolphin named Circe, whose trainer signaled her dissatisfaction at Circe's failure to perform a task on command by taking a few steps back and standing still for a few seconds. When Circe performed as desired, he was rewarded with a piece of fish, but it was known that he did not like the tail end of a fish unless the fins had first been removed. On occasions when his trainer forgetfully tossed him an untrimmed tail, Circe would swim to the far end of the pool and position himself upright in a stock still position. He had appropriated the trainer's signal and was now training the trainer.[8]

One of Aesop's fables tells of a crow who dropped stones into a pitcher to elevate the water to a level where the bird could access it for drinking. When orangutans at the Max Planck Institute for Evolutionary Biology were presented with a clear plastic tube mounted solidly into the floor of their enclosure, at the bottom of which a peanut floated, they at first tried brute force to remove the tempting morsel, biting, hitting, and kicking the tube. But soon (within nine minutes on average) the great apes reasoned a crafty solution. They began taking mouthfuls of water from their drink dispenser and squirting the liquid into the tube until the peanut floated to a point that it could be reached.[9]

No doubt you, like me, have suffered the disappointment of discovering some delicious leftovers that got pushed to the back of the fridge and have since spoiled. Despite these lapses, we know to be more vigilant of items in the fridge than pantry items with a longer shelf life. Studies of birds that cache their food show that they, too, keep closer tabs on the perishables. In a series of ingenious experiments by Nicky Clayton and students at Cambridge University, California scrub jays were provisioned with perishable wax worms (wax moth larvae) and nonperishable peanuts buried in sand-filled storage trays. The jays soon learned that the worms degrade and become unpalatable over time. By giving the birds peanuts to cache (they bury them in the substrate, as squirrels do) on one side of the test arena and worms on the other, the research team was able to observe the jays' foraging choices over time. If the birds monitored the passage of time and knew its effects on the foods, we might predict that they would preferentially dig up and recover worms in the short term, and switch to peanuts following a longer time lapse.[10]

That is exactly what they did. In trials conducted four hours after the birds had cached the nuts and worms, the jays directed more (80 percent) of their recovery inspections toward the worm side of the storage tray. But 120 hours later, by which time worms were decaying, the jays exclusively went for the peanuts. The same preferences occurred even when the food items were removed and replaced with fresh sand just before the 4-hour or 124-hour trials, thus showing that the jays were not merely switching because they could smell that the worms had gone bad, or that they were somehow relying on visual cues.

To be sure that the jays' behavior wasn't due merely to a tendency to more quickly forget worm versus peanut caches (good scientists are diligent about satisfying alternative explanations, however unlikely), they trained another group of jays to expect no worm decay, by replacing the decayed worms with fresh ones just before the 124-hour trials. These "replenish group" birds continued to prefer worms in both 4-hour and 124-hour trials, thereby demonstrating that other birds' preference switch was not a result of greater forgetfulness for worms than for peanuts.

But there's more. Because the chances are greater that a cached food item will be pilfered by another forager if it has been buried longer, the researchers wanted to see if the birds' recovery decisions reflected an awareness of this. By training jays in which all worms had been removed prior to the 124-hour trial, they found that these birds were more inclined to first inspect the peanut tray (29 percent) than were the birds from the replenish group (0 percent). This result shows that the birds appear to learn that the probability that caches will be pilfered increases with the amount of time since caching.

These experiments demonstrate that scrub jays remember at least three aspects or dimensions of their caches: they know what they cached, where they cached it, and when they cached it. Furthermore, they show awareness that some foodstuffs perish sooner than others, and that some are more vulnerable to theft. If only refrigerators were as smart.

A simple but clever study of memory entailed placing a row of eight cardboard "flowers" in the foraging territories of three male rufus hummingbirds. Each of the flowers was a different color and bore a unique pattern. The hummingbirds soon learned that these flowers contained nectar, which was dispensed from a syringe tube mounted behind each bloom. Once the birds were acclimated to the setup (one to two hours), four of the flowers were randomly assigned to be refilled 10 minutes after the bird last visited that flower, and the remainder were refilled following 20 minutes.[11]

At first, all three birds would visit his array of flowers quite regularly and usually at less than 10-minute intervals. But they soon began to revisit the 10-minute flowers sooner than they revisited the 20-minute flowers. Furthermore, while occasional visits occurred sooner than the refill rates, the majority occurred beyond that time. Specifically, over the course of over 1,000 revisits per bird to the 10-minute flowers, Bird 1 averaged 14 minutes, Bird 2 averaged 13 minutes, and Bird 3, 12 minutes. Their average timing of >1,000 visits to the 20-minute flowers was finer: 20 minutes, 21 minutes, and 20 minutes, respectively. Thus, each bird learned that the refill rates differed among the flowers, that the specific refill times were 10 and 20 minutes, and they seem to have remembered specifically which flowers they had emptied recently. Feeding is but one part of these birds' daily activities; they are also constantly occupied with such tasks as guarding their territory against intruding males, displaying to females, and staying alert for potential predators.

Another study, conducted at the University of Düsseldorf, tested the memory patterns of house mice. They placed new objects in a space to see if the mice had the same sort of episodic memory as the rufus hummingbirds. The hypothesis was that if the mice recall objects in their environments, they will tend to ignore them in favor of new objects. In separate trials, individual mice were presented with (a) familiar and unfamiliar objects, (b) familiar objects, some of which had been moved, and (c) familiar objects, some of which had not been encountered as recently as others. In each case, the mice showed more interest in the (a) unfamiliar, (b) moved, and (c) less recently experienced familiar objects. The researchers concluded that house mice do retain a memory of objects in their environment. They know where the objects were, recognize them, and recall their spatial placement.[12] It is plain to see that this skill would be useful to an animal who forages repeatedly in different locations for food sources that spoil with time.

It should be added, though, that these mice were probably at a disadvantage to normal mice. As so-called lab mice, they were reared in traditional lab rodent cages. These small, shoe-box-size units typically contain only a water bottle, hunks of processed rodent chow placed on the roof to be gnawed through the bars, and, if they're lucky, a nest box and nesting material. These conditions thwart strongly motivated behaviors such as foraging, burrowing, nest-building (performed by both sexes), choosing social partners, and—if no nest box is provided—hiding.[13] We've already seen how intensely active mice are. So strong is their drive to forage that caged mice will go searching for hidden food if it is provided, even if chow is freely available in their food hopper—a phenomenon termed "contra-freeloading."[14]

A 2007 study of meadow voles published in *Animal Cognition* concluded that these abundant mouselike rodents from grassy habitats can recall the what, where, and when of a past event. On the first day of the experiment, male voles were introduced to two cages: one contained a female a day away from giving birth; the other held an adult female who was not in any sort of reproductive condition. Female voles reach peak sexual receptivity on the day they deliver their pups.

On day two, male voles were once again introduced to the same two cages, which were now cleaned and empty. They spent significantly more time investigating the cage that had held the expectant mother vole. Thus, the males appear to remember the key information they got the previous day. The researchers used two slightly altered conditions to verify that the pattern was meaningful. In the first of these, males showed no preference when they were reintroduced to clean empty cages just half an hour after the initial introduction (when the pregnant female would not yet be at her peak reproductive potential). In the second case, a peak-receptivity female replaced the pregnant female in the initial encounter. Peak-receptivity in this species lasts only a few hours, so the female would not remain in the peak phase a day later. Once again, males showed no preference for either cage. Thus, meadow voles can perform mental time travel.[15]

Results like this inevitably produce skeptics. A University of Toronto expert suggested that it just could be that the voles were still able to detect some odor on the cleaned cages. If so, it would be a case of a species' impressive mental feat being nullified by its impressive perceptivity. This doubt could easily be satisfied by exchanging cleaned cages on day two with identical new ones.

Clever Bunnies

Rabbits are often dismissed as simpletons with little up top (except for those prominent ears). Rabbit rescuer Andrea Bratt-Frick is often surprised by the alertness of her rabbits. Preferring to have the rabbits' living quarters tidy before their dinnertime, Andrea began training some of them to put a toy in their hay box before giving them their evening salad. Within a week they were getting pretty good at it. The following week, as she began making salads in the kitchen, Amanda started hearing noises in the rabbit cages. She looked out the window and all the rabbits were putting toys in their boxes before she even gave them the command. She figures either they heard or smelled the kitchen preparations, or they had seen what she was doing through the window.

Bratt-Frick is one of a growing number of people who clicker-train domesticated rabbits to negotiate obstacle courses. The training technique uses the same positive reinforcement approach that is so effective with dogs. When a rabbit performs a desired behavior, she is immediately given a small food treat accompanied by the sound of a handheld

Rabbits are intelligent and keenly perceptive. (Photo by Connie Pugh.)

mechanical clicker. The bunnies soon learn to associate the click with the reward and will perform on cue from the clicker. The result is that we have rabbits behaving much like dogs, hopping up and down ramps and leaping over obstacles in a prescribed course.

Frick reports that her rabbits quickly learn new things, and that they sometimes invent novel solutions to problems. Whereas most of the rabbits clear the one- to two-foot-high tiered obstacles with graceful bounds calculated to just clear the top, one of Frick's videos shows a particularly fluffy rabbit named Muffy pausing in front of obstacles, removing a slat with her mouth, then hopping through the gap. Muffy performed the behavior spontaneously, but another rabbit named Mattie didn't. Observing Muffy's shortcut, he also began removing slats. To avoid having the trick spread through the bunny flock like a knock-knock joke through a playground, Frick had to train Muffy and Mattie separately for a while. Social learning in rabbits? You bet.

Misunderstood Fish

The common view that fishes are unfeeling robots with fins is becoming increasingly outdated in the face of emerging studies. Forty years ago, the idea of producing a book devoted to the mental and emotional qualities of fishes was unthinkable. Science wasn't ready to accept, let alone study, fishes in those contexts. Times have changed. In 2005, three fish biologists published an edited volume titled *Fish Cognition and Behavior*. Contributors included fish behavior experts from five continents. As biologist Tony Pitcher notes in his foreword, fishes have had over 60 million years to evolve brains that deal flexibly with diverse underwater environments.[16] Biologist Gordon Burghardt dedicates an entire chapter of his book *The Genesis of Animal Play* to evidence that fish engage in play behavior.

We have many prejudices about fish. To us, they are "lower animals," cold-blooded and machinelike. It's a shallow view considering the sheer diversity of this legion of vertebrates. There are at least 25,000 fish species worldwide, which accounts for more than all the other vertebrate groups (mammals, reptiles, birds, and amphibians) combined. Fish didn't stop evolving when the first lobe-finned member of their kind ventured onto land and established the terrestrial vertebrate lineages. Fishes encompass a diversity of perceptual, mental, emotional, and cultural phenomena.

In their aquatic element, fishes are finely tuned to their surroundings. They have been shown in careful scientific studies to engage in precise discriminations, such as their preference for shoaling (grouping) with fish carrying fewer or no parasites than those with higher parasite loads. They also show social learning both within and between species, predator inspection behavior, and the ability to generalize from one live prey type to another. Even the smallest fishes show awareness of their surroundings. When individual three-spined sticklebacks were presented with two unfamiliar shoals of sticklebacks—one of fishes who were familiar with each other, and the other of mutual strangers—they preferred to swim with the shoal of familiars.[17] Human observers could detect no differences in the swimming behavior of familiar- and stranger-shoals, so it isn't known what the sticklebacks were cueing on. But it's a useful discrimination to make, for it turns out that familiar-shoals locate and consume food more efficiently than do stranger-shoals.[18]

And in case you thought fishes have no sense of sound, they have been shown to discriminate and generalize between different genres of music. In a study that presented recordings of Bach's classical music and the blues of John Lee Hooker to three individual carp, all three subjects learned to distinguish between the two genres, and could generalize from the specific artists to multiple artists presented from each genre (twenty representing blues, nine for classical).[19] Finally, like mice,

Minnows are among several types of fish that show preferences for other familiar individuals. (Photo courtesy of Mike Howell.)

monkeys, and other mammals discussed earlier, fishes also suffer stunted learning capacity when reared in unstimulating environments.

Fishes have a well-developed chemical sense. Water is an excellent medium for diffusing chemicals, and fishes use it both voluntarily and unwittingly. Like the urine signatures of mice, fishes can extract information from each other's chemical secretions. For example, female swordtails select well-fed males based on chemical cues.[20] Many fishes produce chemical cues in response to physical situations such as injury, or to fearful situations. Other fishes cue into this useful information. A fish may release an alarm chemical in the presence of a predator fish (or in response to the odor of one). Other fishes, detecting the chemical, may react appropriately, such as by taking cover or being more vigilant.

Predator inspection involves one or more fishes leaving their shoal, approaching a potential predator, and assessing the likelihood that the predator will attack, before rejoining the shoal. It has been observed in many fish species, including guppies, sticklebacks, minnows, paradise fish, damselfish, bluegill sunfish, and mosquitofish.[21] Predator inspection begins with predator recognition, which in itself is not an innate skill—fishes must learn to recognize specific predatory species as dangerous, and one way they learn is by observing the alarm response (odors and behaviors) of other fishes of their own kind in the presence of a new predator. Because predator inspection is obviously risky, fishes tend not to do it alone. Sticklebacks usually inspect in pairs, and studies find that they prefer to team up with the same inspection partner. Further experiments have shown that a stickleback need observe just four predator inspections by other pairs of fish to be able to remember and favor the better (more cooperative) of the two partners. Guppies also preferred to associate with the better of two inspectors four hours after they had observed them in action. The ability to discriminate individuals within a shoal has now been scientifically demonstrated in several species of fishes.

Another example of monitoring behavior in fishes is known to occur in Siamese fighting fish and green swordtails. Males monitor aggressive interactions between neighboring males and use this information in deciding whether or not to fight another male. Predictably, they are more willing to fight males they have seen lose than to take on a prior winner.[22] Male Siamese fighting fish also alter their threat displays depending on who's watching; if a female is in the audience, he puts more sexuality into his movements by performing more tail-beats and spending more time with his gill-covers erect.[23]

As we learn more about fishes and their mental capacities and sentience, we are beginning to revise our previous dismissive attitude toward them. In Norway, where fishes have been excluded from that nation's Animal Welfare Act since 1974, a new act gives fishes a level of protection on par with other vertebrates (see the "New Days, New Ways" section of Chapter 11). As researchers at the Norwegian School of Veterinary Science point out, because individuals suffer and not species, the ramification of fishes' capacity to feel pain and to suffer are great, given the enormous numbers of fishes humans exploit and kill.[24]

Multitaskers

On a chilly morning while waiting at the bus stop, I saw an example of an animal's ability to process different parcels of information at once. A small spaniel showed increasing agitation from his unfenced yard as a larger spaniel approached on his master's leash. Both dogs took keen interest in each other. The smaller dog assumed an aggressive air typical of dogs restrained on their own territory. He barked energetically and made repeated lunges toward the leashed canine, but he never crossed an imaginary line, and it was clear that the yard was fitted with an invisible electric barrier. After the larger dog was led away (to defecate tactfully on the next lawn down), the smaller one stood watching, took a slight step forward, flinched, and trotted back to safer turf.

The behavior of the little dog is a simple illustration of how animals can and do process different information simultaneously. In this case, the territory owner's fierce desire to approach and make contact with the other dog was overridden by the imperative to avoid an electric shock. This sort of mental multitasking is something that animals, including us, need to do often, as when we keep that dentist appointment in the back of our mind while we finish the grocery shopping.

Here's another example. When starlings were placed in enclosures on natural fields, they foraged less when visual barriers hampered their ability to scan their surroundings. Specifically, the birds spent less time foraging in typical starling fashion—with their heads down. The barriers were designed only to hinder vision when the birds were head-down, and not when they were standing upright. The conclusion was that starlings continue to scan when they are foraging with their heads down. Even though they are prying apart vegetation with their beaks and peering inside for a scurrying invertebrate, these mental multitaskers continue to look out for important information in their ambient surroundings. The researchers also noted that most of the head-up scanning was oriented to other nearby starlings, which suggests that scanning in starlings is mostly about monitoring other flock members.[25]

There is evidence that chimpanzees, like humans, are susceptible to the Stroop effect, which is the conflict that arises in one's brain when one is asked to name the color of ink that a word is printed in, when that word denotes a different color (for example, when the word "green" is written in orange). When presented with a previously learned symbol for a color, but with the color of the symbol itself now changed, the chimpanzee showed poorer accuracy (though not slower speed) in naming the color represented by the symbol. This suggests that chimps can process two different streams of information at once, and detect a conflict between them.[26]

An African antelope known as the dikdik responds adaptively to different predator threats. A dikdik who has spotted a leopard (often before the leopard has spotted the dikdik) will make a distinctive call to alert the leopard. The dikdik may also move toward the leopard, perhaps to ensure she is heard and perhaps to better monitor the cat's movements. Leopards are ambush predators that take their prey by surprise; they will usually not bother to pursue an animal who has detected them first. However, if a dikdik encounters a pursuit predator, such as a pack of African wild dogs, the tiny antelope will quickly and quietly drop to the ground and will often escape detection thanks in part to its camouflaged coloration. If the dogs draw too near, the dikdik will burst away, zigzagging through the dense vegetation with such agility that the heavier predator has trouble following. This behavior requires intimate knowledge of the terrain. But the dikdik doesn't try to outrun its pursuers; it will drop down again suddenly, which often eludes them.[27]

When animals show the hallmarks of having a mind and thinking about things, down tumbles one of the most insidious and destructive ideas of all time: that you need language to think. This was what the seventeenth-century philosopher René Descartes argued when he likened non-human animals to "machines." Descartes further believed that without a mind, one had no soul and did not exist as any sort of being worthy of moral consideration. Descartes was and remains influential, and he is widely credited with helping to foster centuries of cruel and callous treatment of animals in labs, factory farms, circuses, and elsewhere.

Our emphasis on intelligence as the measure of moral standing has a long history. Stephen Jay Gould's book *The Mismeasure of Man* (1981) chronicles the historical penchant of white men to claim intellectual superiority to nonwhite men.[28] Such characterizations were used to justify colonial expansion and the slave trade. Today, racial discrimination is almost universally recognized as immoral. We might ask, What does intelligence matter anyway? As I will argue repeatedly in this book, a species' intelligence is an inadequate basis for moral standing. What matters is not a capacity to think, but a capacity to feel. Or as British philosopher Jeremy Bentham famously said in the eighteenth century, "The question is not, Can they reason? nor, can they talk? but, can they suffer?"[29]

CHAPTER FOUR

with feeling: emotions

Feelings do not fit under a microscope.

—Marc Bekoff[1]

One day, my cat, Mica, perched on the vanity opposite me as I washed my hair in the bath. When I brought my soapy head up from the water, Mica looked at me intently and I said silly cat-talk things to him. "Hello, Mica. Do you like my new hairdo?" The suds give my thinning locks some permanence, so I molded my hair into a peak sticking straight up. I could see from the large mirror behind Mica that I looked satisfyingly ridiculous. Mica's eyes widened slightly but he still looked relaxed. Suddenly, I had an idea: what if I was to sprout horns on my head? I flattened my central peak, and quickly formed two "horns" sticking out and up on each side. I looked at Mica with a grin. His stare intensified, his pupils began to dilate, and he lowered himself gradually into a crouching position. He looked like he was preparing to take flight. I flattened my hair and Mica relaxed and jumped down to come and inspect me. We leaned toward each other and touched noses. Fairly certain that it wasn't merely my soapy head that he was reacting to, I once again sprouted horns, and Mica drew back again apprehensively.

Was Mica recognizing a wild creature in my visage? It would make an interesting study to test cats' responses to familiar faces adorned with various horns, some of which do and do not resemble those of real animals. What isn't much in doubt is that Mica had an emotional reaction to my antics.

Do Animals Have Feelings?

If you've lived with dogs and cats, you know darn well that animals are emotional. But science, to its credit, isn't satisfied by intuition. Science likes to measure things, to test hypotheses and collect data. Until quite recently science wasn't testing hypotheses about animal feelings. From the time Charles Darwin wrote his last book, *The Expression of the Emotions in Man and Animals* (1872) to about the time Neil Armstrong left footprints on the moon nearly a century later (1969), prevailing scientific dogma denied animals their hearts and minds. A nonhuman animal was viewed as merely a responder to external stimuli. The idea that a walrus made decisions, or that a parakeet felt emotions, was considered unscientific.

Fortunately we've moved on, thanks to the work of some curious researchers. No longer is it fashionable to reduce animals to pigeons in a Skinner box pecking at red and green buttons—though that rather primitive scenario is still common in psychology labs. Today, scientists are interested in how animals think, and a growing number are also asking questions about how they are *feeling.* And the public's fascination with animal emotions is illustrated by the success of a 1995 book titled *When Elephants Weep: The Emotional Lives of Animals,* which was the first overview of the subject since Darwin, and which has sold over a million copies.[2]

Emotions evolved as adaptations. They mediate and regulate a wide variety of social encounters among friends and foes. Some biologists speculate that emotions even evolved before consciousness did. It's not a far-fetched idea when you consider the value of a visceral, fear-generated flight from danger. By comparison, individuals who stopped long enough to think, "Hey, this is dangerous, I'd better get out of here," might not be around long to tell the tale. As Marc Bekoff says, the question isn't Do animals have emotions? but rather, What sort of emotions do animals have? My sense is that emotions likely evolved simultaneously with consciousness, for the two serve each other and require a similar quantum leap in mental complexity.

Let's put emotion in Darwinian terms. If an animal is likely to benefit reproductively by feeling and responding to a given emotion, then the case for the existence of that emotion in that creature is made stronger. Fear is one of the most intuitive emotions to this argument, for being afraid of (and avoiding) danger has such obvious and widespread survival value. Most vertebrate animals respond to signs of danger in ways consistent with an experience of fear. They may cringe, flee, freeze, or assume a threatening posture.

It is tempting to think we can use human capacity for feeling as the "gold standard" by which all other species should be measured. But such a view would falsely assume that we share the same sensory capabilities as other creatures, though we already know that is not the case. Nevertheless, we can use our own emotional experiences as a useful template for interpreting the emotions of other beings. Six universally recognized and biologically determined facial expressions of emotion have been documented in humans: anger, disgust, fear, happiness, sadness, and surprise. These so-called primary emotions are widely expressed in other mammals. Studies show that the facial expressions of nonhuman primates share evolutionary origins with humans in both their physiological structure and their social function.[3]

Emotional Physiology

Human parents grieving over a miscarriage or the loss of an infant show elevated levels of glucocorticoids—hormones released by the brain in response to stress. In baboons, higher levels of glucocorticoids have been recorded for a month or more in mothers who have lost an infant. As we may expect in a highly social, long-lived species in which all individuals in the community know all others and form long-term friendships, other members of the baboon troop also show elevated hormones following the loss. But the highest levels occur in the individuals who are most closely related to the infant, and who can be expected to feel the pain of loss the most. This is usually interpreted as a "stress" response, but it seems likely that what these mothers are feeling is grief.[4] The scientists who documented this termed it bereavement. Feelings of loss and bereavement are known in humans to be eased by the social support provided by friends and other loved ones. In baboons, the bereaved individuals compensate for their loss by solidifying their grooming networks. Because grooming is known to lower stress hormones in the participants (both human and nonhuman), it seems fair to regard it in this context as therapy. It helps soothe the baboons' emotional pain and better enables them to "move on."[5]

Can reptiles feel emotions? It is not a question that scientists have shown much burning desire to address. One exception, however, is Canadian physiologist Michel Cabanac, who, despite reptiles' reputation for cold-blooded indifference and independence, found and demonstrated an emotional fever response to human handling in both turtles and lizards.[6] Emotional fever describes a measurable rise in an animal's core body temperature attributable to a psychological rather than physical cause. Impending exams and exciting sports events have been shown to induce emotional fever in humans.[7] Handling by an unfamiliar person elicits in rats a rise in body temperature of 1°C (1.7°F) or more. If the same person handles the rat over several days, the response declines and disappears as the rat develops trust for the handler. If, however, another handler is introduced, emotional fever returns.[8]

Another physiological indicator of emotional response is a change in skin temperature. Negative emotional arousal tends to elicit lower skin temperatures, and the opposite is true of positive emotional arousal. These temperature changes are caused by hormones directing the body to shunt blood away from or toward internal organs by constriction or dilation of blood vessels. Blushing and blanching in embarrassing or shocking circumstances, respectively, are visual manifestations of this phenomenon, and the old saying, "cold hands, warm heart" (a reserved, cool exterior may disguise a kind heart) derives from these changes. Studies show that, in humans, skin temperature can change by up to 2 degrees even when ambient temperatures and genetic factors are held constant. Chimpanzees also show significant decreases in skin temperature when viewing videos of three categories of emotionally negative scenes: other chimps receiving injections, images of darts and needles alone, and another chimp in conflict with veterinarians.[9] When the chimps were required to categorize emotional video scenes—such as favorite food and objects, and veterinarian procedures—according to their positive and negative valence, they spontaneously matched the videos to images of chimpanzee facial expressions according to their shared emotional meaning. These results indicate that chimpanzees process facial expressions emotionally and empathically, as do humans.

Communicating Emotions

While emotions originally evolved to serve the individual experiencing them, as social groups evolved, emotions inevitably took on a role in the way animals communicate. For social animals, the ability to "read" another's emotions allows one to act appropriately, which could mean the difference between a rewarding and a punishing interaction.

Humans are superb at reading the emotions of others. Because there are commonalities in how animals express certain emotional states, we often can read their feelings, too. One way to examine animal emotionality is to measure the consistency with which humans may interpret the possible emotional states of animals. For example, when untrained observers were asked to describe in their own words the behavioral expressions of twenty pigs, there was a high level of agreement across different observers. In the words of the authors of this study, this agreement suggests that their "assessments were based on commonly perceived and systematically applied criteria."[10]

Many factors—especially context, facial expression, and body language—come into play in the interpretation of emotions. We don't have to be taught this stuff—it comes naturally. Marc Bekoff believes that most people without any scientific training can identify emotional responses in animals based on muscle tone, posture, gait, facial expression, eye size and gaze, vocalizations, and in some cases odors.[11] Rarely does any one of these cues occur in isolation; usually, our interpretation of what an animal is feeling is based on a gestalt or "package" of several cues, as well as the context (such as the physical setting or situation) in which the animal is behaving.

Even when we lack contextual information, we still seem to have an intuitive grasp of some species' emotional states. Consider the experiments of Nicholas Nicastro of Cornell University. Nicastro recorded one hundred meows from a dozen cats in different situations, including the "distressed" meows of cats placed in a car and the "moody" meows of cats groomed beyond their patience. When he played these meows to people and asked them to rate each meow on a scale of pleasantness, and to another group of people asked to rate them according to their urgency, both groups showed a high level of rating accuracy. Meows rated as pleasant tended to be short and high frequency (ME-ow), whereas urgent meows tended to be long and lower-frequency, changing to a higher pitch at the end: me-OWww![12]

Gratitude is a social emotion because it is felt, and usually expressed, toward another being. Primatologist Frans de Waal of Emory University describes a touching example of gratitude expressed by a chimpanzee. Roosje was an infant chimpanzee born in Holland's Arnhem Zoo to a deaf mother who couldn't properly care for him because she couldn't hear his cries for help or attention. Fearing for the little chimp's welfare, de Waal decided reluctantly to move Roosje to safety. Kuif, another chimp in the same colony, had lost more than one of her own infants and had suffered deep depression each time, marked by rocking, self-clutching, refusing food, and heart-wrenching screams. De Waal decided to train Kuif to bottle-feed Roosje through the bars of the chimps' compound. Kuif took well to this, and was eventually given Roosje to bottle-feed. Chimpanzees tend to disapprove of taking someone else's infant, and Kuif glanced between Roosje and de Waal, kissing each, as if asking permission. She was the most caring and protective mother that could have been hoped for. Up to that time, Kuif had had a rather neutral relationship with de Waal, but from that day onward she showered him with the utmost affection whenever he would show his face. Three decades later, Kuif's gratitude remained undiminished.[13]

When a female humpback whale became entangled in the ropes of crab traps near the Farallon Islands (aka the Farallones) off the coast of San Francisco in December 2005, several divers wielding knives dove into the water in a rescue attempt. Rope was wrapped at least four times around the humpback's tail, midriff, and the left front flippers, and there was a line in the whale's mouth—yet, the whale remained calm while the divers cut through the ropes. "When I was cutting the line going through the mouth, its eye was there winking at me, watching me," said James Moskito, one of the rescue divers. "It was an epic moment of my life." Once free, the whale didn't swim away but approached and nuzzled each diver.[14]

An outpouring of gratitude signals something profound. It shows that the animal values his or her life. It also suggests feelings of relief from pain and fear and the fulfillment of a fundamental desire for freedom.

Ambient States

Many emotions are fleeting, and we think of them as temporary states. But we also experience longer-term feelings. We might refer to these as moods or dispositions. They may be changed by a traumatic experience, such as the death of a loved one that may set us back for a prolonged period. I refer to these long-term feelings as our ambient emotional states. Normally your emotional tenor is lower if you are stressed or worried about something, and it rises again when you have resolved the source of your anxiety.

Do other animals have ambient emotional states? We already know that some of the variation in emotional states among individuals may derive from personality differences, which are scientifically well documented in many animals, from hyenas to guppies to octopuses. Demonstrating changes in ambient emotional states is more challenging, but recently an ingenious approach has been used to address the question.

In a study conducted at Newcastle University in England, five-year-old European starlings were housed socially for ten days either in cages enriched with branches and water baths, alone or in smaller, barren cages. Next, the birds were trained to forage by plucking the lids from dishes in which a worm had been placed. During this training period, all birds soon learned that dishes with white lids contained tasty worms, whereas dishes with dark gray lids harbored unpalatable, quinine-flavored worms. Birds from both the environmentally enriched and the environmentally impoverished groups soon ceased bothering to flip dark gray lids. But when the experimenters began presenting the starlings with ambiguous dishes—lidded with lighter shades of gray—they found that only the enriched birds were likely to flip over the lids and sample the worm inside. Birds who had been recently switched from enriched to impoverished cages showed the most pessimistic response.[15] The starlings' responses mirror those of humans suffering from depression or anxiety, who are also known to have more negative expectations and judgments about events and to interpret ambiguous stimuli unfavorably.

These remarkable results suggest that enriched starlings are more optimistic than impoverished and presumably less happy ones. It appears that these animals, like humans, can suffer from depressed morale. Furthermore, it shows that life for a bird can go well or ill, and that an individual's emotional state has duration over time beyond the fleeting emotions of a given moment.

It is well known that animals respond poorly to living conditions that do not stimulate them mentally or physically. Rats, mice, monkeys, and other mammals confined for long periods in laboratory cages where they have little or no opportunity to engage in such natural behaviors as foraging, hiding, nest-building, or choosing social partners develop neurotic behaviors. Termed "stereotypies," these behaviors involve repetitive, functionless actions sometimes performed for hours on end. Rodents, for example, will dig for hours at the corners of their cages, gnaw at the bars, or perform repeated somersaults. These "behavioral stereotypies" are estimated to afflict about half of the 100 million mice currently used in laboratory tests and experiments in the United States.[16] Monkeys chronically confined to the boredom, stress, and social isolation of laboratory cages perform a wide range of abnormal, disturbing behaviors such as eating or smearing their own excrement, pulling or plucking their hair, slapping themselves, and self-biting that can cause serious, even fatal injury. Severely psychotic human patients display similar behaviors. If you've seen the repetitive pacing of caged big cats (and many other smaller animals) at the zoo, you've witnessed behavioral stereotypies.

Birds also develop these neuroses. Caged parrots become obsessive feather-pluckers and may render themselves completely bare; they may also perform repetitive nodding or twirling head motions. In the case of the starlings, the research team had previously found that birds housed for one week with the same enrichments used in the current study showed fewer repetitive behavior patterns and lower levels of corticosterone (a stress hormone) than starlings housed in standard, unenriched cages.[17]

Variations on the starling study design have shown the same pattern of optimistic and pessimistic outlooks in rats and pigs. Faced with between zero and two negative daily interventions (e.g., an unfamiliar cage, a tilted cage, a strange rat, temporary reversal of the light/dark cycle, bedding left damp), rats tended to show fewer responses to the sound of pure tones of ambiguous length that hinted at a positive outcome than did rats kept in predictable housing.[18]

Another method used to demonstrate ambient emotional states in animals has been to monitor their use of anxiety-relieving drugs.

Once again, the patterns reflect individuals with feelings. Rats housed in enriched environments self-administer amphetamine (a mood-altering drug in humans) less than do rats housed in impoverished cages; given the choice of a water dispenser containing the drug, or an unadulterated dispenser, they favor the former one. It's a finding that suggests that the impoverished rats are feeling down (though scientists prefer the term "stressed").[19] Similarly, mice housed in standard (boring) cages drank significantly more water that contained an anxiety-reducing drug than did mice in cages enriched with a nest box, a running-wheel, two cardboard tubes, and two sheets of absorbent paper for nesting material.[20]

There are innumerable descriptions of "tension in the air" in primate societies during periods, for instance, when dominant individuals are facing challenges from others rising through the hierarchy. Rival males often perform violent displays, and it understandably puts everyone on edge. One of the first activities to go is play, which is a dispensable activity that is most prevalent when all is well and good. A prolonged cessation of play usually signals a change in emotional state. The phenomenon has also been recorded in birds. One of the few occasions when playful activity was not observed in a long-term study of Arabian babblers was during two days when a dominant male and two closely ranked subdominant birds were vying for leadership of the group.[21] It was as if every individual in the group was aware of the unrest, for normally playing consumes an hour or more of a babblers' time each day.

What do these ambient emotional states say about the lives of animals? It says that they aren't merely alive, but that they have lives. Life plays out like a film with a narrative rather than a succession of still photographs. Some emotional experiences are intense and short-lived, while others are prolonged and form part of the daily tenor of our lives and theirs. Even during a period of emotional depression people experience moments of mirth or delight. Waves continue to wash ashore whether the tide is coming in or going out.

Taking an Interest

Have you ever been in a situation where there are throngs of people and just two dogs, and as soon as one dog spots the other her attention is riveted on the other canine? I'm always bemused by this because it indicates that from a dog's perspective, another dog is infinitely more interesting than a human. I suspect this tendency to favor one's own species is widespread for adaptive reasons, though there are bound to be common exceptions. Even a parasite, whose lifestyle hinges on gravitating toward a host species, nevertheless has to fraternize with other members of its kind to reproduce (though there are some parasitic worms that save themselves the bother by hitching up in a permanent state of copulation.)

Like dogs, elephants show greatest interest in others of their kind. Their fascination with bones occasionally extends to the bones of other species, but their closest attention appears to be directed toward the bones of other elephants. To test this hypothesis, Karen McComb and colleagues with the Amboseli Elephant Research Project in Kenya conducted experiments in which elephants were presented with ivory tusks and bones of elephants along with the remains of other animals and inanimate objects. Their findings indeed showed that elephants spend significantly greater time exploring elephant remains than inanimate objects or the remains of other large herbivores.[22]

Now that cognition and emotion are more readily ascribed by scientists to other animals, scientists are beginning to acknowledge the depths of feeling that may accompany these behaviors. Combining GPS tracking data, long-term association records of who spends time with whom, and direct observations, renowned elephant researcher Iain Douglas-Hamilton and colleagues describe elephants showing compassionate behavior toward others in distress. In Kenya's Samburu Reserve, Eleanor, the matriarch of a family unit called the First Ladies, became gravely ill and fell to the ground. Grace, the matriarch of another family called the Virtues, immediately went to her aid. Seeing Eleanor down, Grace ran over to her with tail raised and temporal glands streaming secretions, sniffed and touched Eleanor with her trunk and foot, then used her tusks to help lift Eleanor to her feet. Grace appeared stressed and continued to try to lift Eleanor with her tusks and foot after she collapsed again. During the next week following Eleanor's death, elephants from five family units visited her body. Douglas-Hamilton's team also documented general awareness and curiosity about death, and that these behaviors are directed both toward kin and toward nonrelated individuals. The study authors conclude that elephants show compassion toward one another and have an awareness and interest in death.[23] Might it be that elephants have a sense of their own mortality?

Experiencing Loss

The response of elephants to elephant bones might be just curiosity, though it also hints at feelings of sadness. It is fairly clear that elephants recognize the bones of their own kind; otherwise why would they spend more time inspecting them than the bones of, say, a hippo or a rhino?

There is also anecdotal evidence that elephants can feel regret. If animals can experience regret, then elephants would be a good candidate species because they are a long-lived, social species whose individuals remember the actions of others. A tragic incident at the Elephant Sanctuary in Hohenwald, Tennessee, suggests feelings of regret in Winkie, a forty-year-old elephant. In the summer of 2006, during a routine inspection by her two human caretakers, Winkie suddenly struck one of them, thirty-six-year-old Joanna Burke, across the chest and face with her trunk. Burke was sent flying to the ground by the blow, and Winkie stepped on her, killing her instantly.

Winkie was withdrawn for weeks following the incident. Perhaps she was mourning the loss of her caretaker. Winkie's life prior to her arrival at the Elephant Sanctuary in 2000 had been one of confinement in zoos. Her early history is uncertain, but like many captive elephants her age, she may have suffered the emotional trauma of witnessing her entire family being brought down by rifle bullets. Thousands of such "culling" operations have been carried out in parts of Africa as expanding populations (more often human than elephant) create conflict over diminishing habitats. In May 2008, South Africa announced that it was going to resume elephant culling operations, after a thirteen-year moratorium.[24]

Animal friendships are not confined to their own species, and we may expect that loss can be felt for other species, too. In November 2007, four members of the army of Zimbabwe, wearing camouflage uniforms and carrying automatic assault rifles, tied up three guards assigned to each of three black rhinoceroses, then opened fire on the animals, killing them. The three huge beasts—Amber, DJ, and

Sprinter—were members of a small population lovingly reared and bred by the Travers family, who own the Imire Safari Ranch where the crime took place. Six weeks earlier the rhinos had been dehorned to make them less attractive to poachers, but the murderers still tried to hack out the few centimeters of new horn growth on one of the victims before being frightened off.

There were other victims of this senseless, violent crime. Seven-week-old calf Tatenda was left orphaned, and Amber was due to give birth to her calf the following week. To those mourning the loss must be included three elephants, Mundebvu, Makavusi, and Toto. These elephants often walked around with the three rhinos. They all knew each other and at night the elephants slept just outside the compound where the rhinos were kept for protection. They must have been very aware of the shots and screams of the rhinos, according to a Travers family member following the attack. Two days after the incident, the rhinos' three elephant companions were taken to their burial spot. According to witnesses, the elephants reacted very strongly to the site, passing sticks to each other with tears running down their faces. Mundebvu (herself pregnant), dug down over three feet into the earth, apparently trying to reach her fallen companion, all the while letting out screams and shrieks.[25]

Elephants have much to cry and be bitter about. Like so many of Africa's charismatic fauna (and that of all the other continents), populations have been reduced to fragments of their former scale. Ethologist and author Gareth Patterson took me for a walk through the forests near Knysna, South Africa, where he has been studying the relict elephant population that took shelter there and has survived despite prolonged persecution by European settlers and their subsequent elimination from most of South Africa. Human signs nailed to trees demarcating this as elephant territory are often damaged by the elephants, who seem to take exception to such human markings. Gareth showed me a large metal signpost weighing about 10 kg that had been torn down and gored. Ironically, the sign read: DANGER! HARVESTING IN PROGRESS.

In 2005, journalist Kathleen Stachowski witnessed the shooting of an adult male bison near Yellowstone National Park, Montana. The targeted animal—a descendant of the fortunate twenty-three Yellowstone refugees who escaped the massive bison genocide of the 1870s—sat

An elephant's comment on deforestation—damaged logging sign from Knysna, South Africa. The hand holding sign at upper left indicates scale. (Photo by the author.)

calmly in the grass among four of his comrades. The first shot brought all four animals to their feet. It took two more shots before the victim fell down again, at which point the remaining three gathered around him. Stachowski records what followed:

> The carriage of their tails registered distress. One, in particular, seemed especially anguished; he pawed the motionless shoulder as if to rouse him. Getting no response, he nudged the body with his head then with the shank of his horn. Again and again he nudged and butted . . .[26]

It may not be an overstatement to say that a zebra stallion feels grief at the loss of his son. Biologist Cynthia Moss, who has studied African wildlife for over thirty years and has written several books on the animals' behavior, refers to the "extremely strong personal bonds" between them when an older stallion returned several times to his dead four-and-a-half-year-old son and repeatedly tried to rouse him. Later he left his family group and searched the heard, calling for six hours.[27]

Susceptible Psyches

After a visit to the veterinarian, our two-year-old male cat Mica (introduced at the beginning of this chapter) was traumatized. He didn't eat for forty hours, and he looked around corners nervously as if the cat-carrier was about to reappear. His sensitivity to the event is all the more remarkable considering that it wasn't even Mica who had been to the vet—his sister Megan had. Mica hissed at her for days.

I resisted taking Mica to the vet for routine vaccinations and checkups for some time after this. Then, when he developed a urinary disorder, I was compelled to pack him into the cat carrier for the ten-minute drive. I felt guilty shoving him inside the carrier and latching the door. Mica wheeled around, spewing a tirade of yowls, growls, and hisses I wouldn't have known he had in him. I felt such pity as he lunged at the cage door, and guilt for the control I wielded against his will. Ninety minutes later, I gratefully placed the carrier on the floor, opened the door, and Mica, following a moment's hesitation, shot out. This time he recovered much sooner, and it was Megan who consigned herself to the upstairs floor, but thankfully only for a few hours.

I find such incidents sobering, for they illustrate just how acutely sensitive and emotional animals can be. In light of them, it is remarkable that the existence of emotions in animals has been so much in dispute when they express their feelings so markedly for anyone paying attention. Animals typically recover from traumatic episodes, but a growing animal psychiatry industry attests to animals' vulnerability to a range of psychological conditions. Dogs are prescribed antidepressants for the treatment of separation anxiety, and behavioral therapy is prescribed to cats suffering from "territorial disorders" following a move. In a paper on the treatment of emotional distress and behavioral disorders in cats and dogs, veterinarians Amy Marder and Michelle Posage list twenty-four commonly prescribed drugs.[28]

As long-lived, intensely social animals, elephants are vulnerable to the same sorts of long-term psychological conditions that may afflict humans who have suffered mental or physical trauma. Gay Bradshaw, an animal psychology researcher and founder of the International Association for Animal Trauma and Recovery (IAATR), has been studying post-traumatic stress disorder (PTSD) in elephants. Many elephants have been victimized by poaching and so-called elephant culls in which entire family groups are gunned down, leaving only the youngest calves to be raised by compassionate humans. Growing up with the traumatic memories of violence and loss, and exacerbated by the lack of nurturing that only one's own mother can provide, these orphans show the classic symptoms described for human PTSD patients: sleep disorders, reexperiencing (including what appear to be nightmares), loss of appetite, irritability, and hyperaggression.[29] Some of these victims grow up to be dangerous rogues. Three adolescent males, all victims of culling campaigns, went on a killing spree, leaving over sixty rhinos dead. At another heavily affected African park, nearly 90 percent of all male elephant deaths were caused by other male elephants, compared with 6 percent in relatively unstressed elephant communities.[30]

The international trade in chimpanzees for zoos, pets, and laboratory subjects has inflicted psychological trauma on many chimpanzees, too (at the time of this writing, the United States was using over 1,200 chimps in laboratory experiments). On a visit to a sanctuary for chimpanzees "retired" from laboratory research, I saw individuals with a variety of bizarre behaviors, including one male who needed to always be in contact with a wall of his outdoor compound. Several scientists have begun to describe in these animals symptoms formally associated with PTSD in humans, including avoidance, spontaneous angry outbursts, failure to socialize, and sleep difficulties.[31]

An incident at Fauna Foundation, a chimpanzee retirement sanctuary for chimps formerly used in laboratory experiments, illustrates the power of memory of past trauma to arouse chimpanzees. When a shipment of materials was loaded into an unused trolley and pulled past the chimpanzees there, two of the chimps—Tom and Pablo—simultaneously let out a piercing shriek. At that point, all fifteen chimps lunged forward and clung to the bars of their enclosure, rocking back and forth while screaming and staring at the trolley. It later came to light that this trolley had been used to transport unconscious chimpanzees from their cages to the surgery room in the same laboratory from which Tom and Pablo had come two years earlier.[32] The incident also bears some of the hallmarks of PTSD, including triggers and fear. As this syndrome is studied more closely in nonhumans, it seems likely that it will be found to be widespread for them, especially as they are often subjected to treatment that is painful, fearful, and unjust.

As with their capacity for physical sensations, animals' emotional sensitivity compares to our own. It is no more scientific to assert that a nonhuman lacks feelings or is emotionally duller than us than it is to say that an animal has a poorer sense of taste or touch. Humans may experience certain emotions that other species lack, and some emotions more intensely, but the obverse is also probably true. The challenge for the pioneering scientists studying animal emotions is to devise creative, ethical methods to reveal what the animals are feeling. As perceptive, cognitive beings, animals can express preferences, which is a window onto their feelings. Recent studies with cows find that the amount of eye-white visible in the animal's eye increases during temporary separation from their calves. It also corresponds to their frustration when food is temporarily withheld, and to contentedness when it is provided.[33]

Behavior and context also provide good clues to how animals are feeling. Most people who are familiar with domesticated dogs and cats can recognize fear, excitement, distress, contentment, and other emotions in these animals. With less familiar species we must turn to experts who know their behavior—preferably in the wild, where it is uncontaminated by artificial constraints of captivity. In baboons, for example, a tail held in the air is the equivalent of a dog's tail between the legs.

The important thing for humans to recognize is that animals do indeed have feelings, and that to the animals these are every bit as important as our own feelings are to us. As we awaken to animals' emotional sensitivity, we'll become more attuned and, I hope, more sympathetic, toward their emotional vulnerabilities and needs.

CHAPTER FIVE

knowing it: awareness

"None of the animals has conviction... none has reason."
Aristotle (384–322 BCE)

At the harbor in Hobart, Tasmania, in July 2007, I watched an interaction between a Tasmanian raven and a subadult Pacific gull. The gull, huge even by raven standards, stood atop a broad, flat stump with a large gristly lump of dead sea creature in his bill. The raven, perched on an adjoining wooden rail, was interested in getting a piece of the spoils for herself. Several times she stalked up and strategically tried to pull away all or part of the morsel. At one point she succeeded in getting hold of a dangling piece of the gristle. During the ensuing tug-of-war, the gull vocalized and held on. The equally stubborn gristle refused to break, and the gull managed to win the contest. As soon as the gull had successfully swallowed the prized object (which to me looked thoroughly disgusting), the raven hopped onto the stump. As both birds poked around for any remaining tidbits, they tolerated each other's close proximity. The gull then suddenly took a quick step toward the raven, causing the black bird to fly away.

This was entertaining for an inveterate animal watcher like me. But it was nothing especially rare or noteworthy. Opportunistic birds like ravens and gulls often pit wits against one another in pursuit of food or other desired goods. What is noteworthy in the context of this chapter

As opportunists, gulls are keenly aware of what's going on around them. (Photo by the author.)

is the manifest awareness of the participants. The timing and tenor of each bird's movements were those of a conscious, flexible being. The raven's day, like the gull's, is lived consciously. Though we may never know intimately what a gull's or a raven's umwelt is like, we can still look on them from our human perspective and experience their worlds empathically.

What does it mean to be aware? In the preceding chapters we've seen that animals are keenly attuned to their surroundings, that they possess intellect suited to their ecological station, and that they have a range of emotions. All of these capacities are relevant to awareness. To be aware is to be vigilant and alert to one's surroundings. It is also to have a mind.

Here is an example from a study of elephants that will give an idea of their awareness. Individual samples of female elephant urine mixed with earth were placed in the paths of thirty-six different elephant families, and the reactions of the passing elephants were observed. Elephants showed little interest if the urine sample was from an elephant outside their group, but stopped to reach with their trunk if the sample was from a family member. The response was noticeably heightened if the fresh urine was from a family member who was walking farther behind. Urine from a family member who was far away and not in the herd also elicited a heightened response. The researchers described the reaction as one of "surprise" because the pachyderms' experience didn't fit their reality. Based on their observations, the team estimated that African elephants are able to keep tabs on at least seventeen females and as many as thirty individuals of both sexes at any given time. These elephants keep mental tabs on who's who and who's where.[1]

Trying to explain the physical origins of awareness is regarded as one of science's most vexing problems. But the fact that consciousness evolved need not surprise anyone. Once creatures began to move about it was, arguably, only a matter of time before consciousness followed. Awareness is enormously adaptive. It enables the organism to make sound choices. In a complex, changing world, that's very useful.

Another's Perspective

If I were to list the most convincing ways that one organism can demonstrate awareness, near the top of my list would be the ability to view a situation from the perceptual perspective of another. Scientists explain this sort of awareness in terms of a *theory of mind.* To have a theory of mind is to be aware that another individual also has a mind.

There really can no longer be any legitimate doubt that great apes, our fellow primates, possess a degree of awareness on par with our own. If you've watched chimps, orangutans, or gorillas—most likely in the relative confines of a zoo—you may have noticed an air of reflective dignity and personhood in their mien. Studies of chimpanzees and other great apes have repeatedly shown that they attribute conscious awareness to others of their kind, and to humans. They use this information to devise what could be called effective social-cognitive strategies.[2]

All four species of great ape are known, from both wild observations and carefully designed captive studies, to follow the gaze of another to distant objects and around obstacles. Following another's gaze is adaptive because its direction often predicts what an individual is attending to and its future actions. The apes also reliably put themselves in positions from which they can see what a human was gazing at behind a barrier.[3]

Another series of experiments aimed to see how one chimpanzee might adjust his behavior according to his awareness of what another chimp knows or doesn't know. A subordinate and a dominant chimpanzee peered through a window on opposite sides of a room while a piece of food was strategically placed (baited) somewhere in the room, then they competed for the food after the doors were opened. The subordinate chimp could always see the baiting, but opaque barriers in the room allowed some baitings to be made out of view of the dominant chimp. The subordinate could also monitor the visual access of the dominant.

When baitings were out of view of the dominant chimp, subordinates preferred to approach and retrieve food; they did not do so when the dominant had seen the baiting, because the dominant would know where it was and claim it for him- or herself. However, if a dominant individual who witnessed a baiting was replaced with another dominant individual who had not witnessed it, subordinates went straight to the food, thus demonstrating the ability to keep track of precisely who has witnessed what.[4]

Chimpanzees show knowledge of another's sensory awareness, and they use the presence of the eyes as a cue that their own visual signals will be effective. When chimpanzees are shown the location of food locked inside a box but are not provided a key to open it, they simply lead the way to the box when a human enters the room. But when the human enters with several "blindfolds" wrapped around different parts of his or her face (e.g., forehead, nose, mouth), they recognize the handicap. Chimps have shown two solutions to this situation: either they take the human's hand and physically lead him or her to the box, or they remove only the blindfold covering the helper's eyes, then lead him or her to the box.[5] Chimps also demonstrate attribution (recognizing consciousness in another) by producing more visual signals when a partner's eyes are uncovered, but more acoustic signals when the other's eyes are closed.[6]

Another illustration of the awareness of another's agency is when chimps and gorillas divert a competitor's attention from a hidden source of food when they come on the scene. To enhance the ruse, they may also take to grooming or some other activity. Only when the competitor has left the scene do they again look at the food and retrieve it.[7] This example by no means suggests that chimps are inclined to selfishness, as we'll see later on.

Captive chimpanzees will gesture, vocalize, and display more often when a nearby human is carrying a tool they could use to access food. Chimps also respond differently based on their perception of how engaged the human appears to be at that moment. They gesture more animatedly if the human seems inattentive, and when given the wrong tool persist in their communicative efforts.[8]

If you've spent time with dogs you'll know how attentive they are to our activities and moods. As highly social animals, dogs have evolved to be expressive and perceptive of the expressions of others. A recent study also shows that they, like us (and at least some other animals), can discriminate between "attentive" and "inattentive" humans. In a fetch-game situation, dogs invariably brought the object to the front of the person, even if he or she had turned around to face away from the dog. Dogs presented with two humans eating food prefer to beg from the one who is not blindfolded; they also show more hesitation in approaching a blindfolded guardian than when the guardian's eyes are visible.[9]

Another recent study shows that dogs appear to form mental images of people's faces. Scientists placed twenty-eight dogs in front of a computer monitor blocked by an opaque screen, then played a recording of the dog's human guardian or a stranger saying the dog's name five times through speakers in the monitor. Finally, the screen was removed to reveal either the face of the dog's human companion or a stranger's face. The dogs' reactions were videotaped. Naturally, the dogs were attentive to the sound of their name, and they typically stared about 6 seconds at the face after the screen was removed. But they spent significantly more time gazing at a strange face after they had heard the familiar voice of their guardian. That they paused for an extra second or two suggests that they realized something was amiss. The conclusion drawn is that dogs form a picture in their mind, and that they can think about it and make predictions based on that picture. And, like us, they are puzzled when what they see or hear doesn't jibe with what they were expecting.[10]

Attributing awareness to another is only one way that an animal may demonstrate a theory of mind. Another way is recognition of one's self. If, when presented with a mirror, an animal recognizes that the reflection belongs to him or herself, then the animal is considered to have shown an awareness of him or herself. The test is performed by surreptitiously applying a mark to part of the animal's body (e.g., the forehead) that is only visible in the mirror. If, on seeing their reflection, an animal responds to the mark by either by touching it or trying to remove it from their body, they pass the test.

Until recently, the only creatures to conclusively show mirror self-recognition were mammals: humans, great apes, dolphins, and elephants. Then, in August 2008, a team of German researchers demonstrated this advanced cognitive feat by secretly placing a colored adhesive dot on the throats of several magpies. On seeing their reflection, some birds tried to remove the dot by scratching with their feet; others tried to pull it off with their beak. Black dots, which were camouflaged against the magpies' black neck feathers, were ignored.[11] The magpie study is an important advance in what we know of the evolution of self-awareness because it shows this capacity has evolved independently in a separate lineage. The special significance of this finding is that brains of birds have evolved differently from those of mammals. Lacking a neocortex—that prominent, often convoluted part of the mammalian brain—birds' brains have instead undergone proliferation in the paleocortex, which in mammals is not responsible for cognition. In 2005, a team of experts announced a makeover in the naming of bird brain parts to reflect the recognition that birds' well-developed paleocortex endows them with cognitive skills akin to mammals.

As primatologists Dorothy Cheney and Robert Seyfarth of the University of Pennsylvania note in their book *Baboon Metaphysics: The Evolution of a Social Mind,* proving that an animal has a theory of mind is very difficult because in most cases their behavior can be explained by relatively simple behavioral contingencies. Dogs who slink guiltily after tipping over the garbage may also do so if another dog did it, so the dog may merely have learned that when garbage is tipped over, people begin to shout. The failure of monkeys to "pass" the well-known mirror self-recognition test has fostered the belief that these animals have no concept of self. While the accuracy of that conclusion can be debated (failure at mirror self-recognition needn't mean lack of self-awareness), a recent study of macaques suggests that monkeys understand other beings as agents with their own perspectives and intentions. Macaques held captive at the University of Parma, Italy, were shown a goal-directed task in which a woman reached over a tall obstacle to pick up a visible toy resting on the other side. On average, this rather predictable scene caused the monkeys to gaze at the woman's face for only 7 milliseconds. However, their average gaze increased to around 18 milliseconds when the woman's behavior appeared irrational—that is, when she performed the same movements with no barrier present. In human children presented with the same scenes, prolonged gazing is thought to indicate that the children have some understanding that the observed person is a rational being with her own intentions, and the same conclusion is drawn from the macaque study. When observing the gazes and actions of other animals, brain activity in monkeys and humans is similar, which provides further support that there is some form of attribution going on.[12]

Theory of mind can also pertain to knowledge of one's own knowledge, or metacognition. For example, when rhesus monkeys are given the option to refuse a memory test, they almost always opt to refuse it if they are (apparently) unsure what they had been shown earlier, or if they had been shown a blank screen. This implies that the monkeys know that they don't know something.[13]

Here's an experiment from the University of Georgia that set out to examine whether rats also have metacognition. Rats were rewarded for pressing the correct lever assigned to either a short (2 to 3.6 seconds) or a long (4.4 to 8 seconds) noise played back to them. Before they could press a lever, however, they were required to poke their noses into one of two holes, one of which voided the current trial and advanced to the next one, and the other of which communicated that the rat was going to press one of the levers. Food pellets were dispensed depending on what choice the rat made: a small reward of pellets for choosing to advance to a new trial, a large reward of pellets for a correct lever choice, and no pellets for an incorrect lever choice.[14] As you can see, just learning the experimental set-up requires some cognition on the part of the rats. The next step was even more revealing.

If the noise was obviously long (e.g., 6 seconds or more), the rats almost always elected to take the test, which usually resulted in their getting 6 pellets for a correct answer. But if the discrimination was difficult (e.g., a 4.4 second noise), the rats usually elected to void the trial, earning a smaller reward (3 pellets) in the process. In trials in which rats were not given the option to decline, they showed the lowest accuracy on the most difficult discriminations. Thus, rats appear capable of judging whether they have enough information to pass a test. They know what they know. They demonstrate metacognition.

Monitoring Others

When it comes to being able to understand human behavior, no mammal comes even close to the dog, says Juliane Kaminski of the Max Planck Institute. Dogs can do things that were long thought only the province of humans. They know they can more readily solve a problem—such as trying to get at food locked in a tightly-sealed container—by sitting down next to a human and staring at them than by gnawing at the container, as a wolf would do.[15] Wolves are no less intelligent than dogs; they just haven't had the advantage of cohabiting with us for a few millennia.

A 2007 study of domestic dogs astonished scientists for what it demonstrates about dogs' awareness and powers of deduction. It had long been known that human infants as young as fourteen months old distinguish an action that is necessary from a logically unnecessary one. For instance, if the infant observes an adult using her forehead to flick a light switch on or off when her hands are empty, the infant will imitate the eccentric method, using his or her forehead to flick the switch. But if the adult performs the same action while clutching a blanket in her hands, infants will use their hands to flick the switch. Presumably, infants can reason that the adult used her forehead because she had no choice. Monkeys show a similar sort of awareness. If a human touches a container with the elbow when her hands are full, monkeys pay much closer attention than if the person touches the container with the elbow when her hands are empty. The animals understand, as a human would, that it is rational to use one's elbow if one's hands are full, and not so if they are empty.[16]

In the dog study, a female border collie named Guinness was trained to press down on a hanging bar to release kibble from a box by using either her paw or her mouth. Dogs normally favor their mouths over their paws to perform such tasks, and a first group of 14 mixed-breed dogs who had learned to use the bar, but who had not observed Guinness, pressed it down with their mouths about 85 percent of the time. A second group of 21 dogs also usually used their mouths (79 percent of the time) to press the bar after observing Guinness pressing the bar when she had a ball in her mouth. But for a third group of 19 dogs, after observing Guinness pressing the bar with her paw even though her mouth was empty, they imitated Guinness, pressing the bar with their paws in 83 percent of the trials.[17]

The most sensible interpretation is that the dogs recognized when Guinness's mouth was busy, necessitating the deployment of a paw to perform a task. More generally, the study suggests that dogs can put themselves inside the head of another dog and make rather sophisticated decisions based on what they witness.

At a local sanctuary for rescued farm animals, I helped dispense medicines to the pigs, many of whom have arthritis. In modern-day meat production, pigs have been selectively bred to grow fast and are sent to slaughter when they are just six months old and weighing about 250 pounds. At the sanctuary, they grow up to three times that size. Bridget, a sanctuary employee, and I squished three blue ibuprofen pills into a third of a banana (peel and all) then plopped each of the spiked morsels into a bucket. Bridget does this daily for the pigs and was happy to show me how to do it. As we wandered among the pigs, laid out and dozing on thick piles of hay in their barn on this freezing February morning, I noticed that pigs knew the routine. Those who are given their daily meds stirred and looked up as I approached, and many of them opened their mouths before I even got there. Others didn't budge. Bridget explained that these latter pigs weren't on meds. I concluded that the arthritic pigs anticipated their treat, and the fitter ones didn't bother stirring for something that they knew wasn't forthcoming.

Squirrels don't accept ibuprofen in a banana, but they do adjust their predator monitoring behavior according to the size of their food. When they encounter a small piece of food such as a sunflower seed that can be eaten quickly, they eat it on the spot, but if the food is larger and requires more time and handling to eat, they will move to a location where they can watch for predators. This was discovered by placing different-size foods inside a 2.5-foot-high frame that blocked the squirrels' peripheral views of the urban park where they were observed.[18]

Many animals show audience effects, whereby they change their behavior depending on who is watching them. It's another form of awareness, and it's one that fishes are capable of. When a male Atlantic molly is given the choice between a larger and smaller female of his species, he spends more time near the larger female. This is adaptive, for larger mollies tend to produce more eggs, which means more offspring for a male who mates with her. But if another male is introduced into the mix, the first male spends less time near the initially preferred female and more time near the other, smaller female. This shift in proximity also occurs when a male of a different species of molly is introduced, but it is much weaker.[19] It isn't clear whether this is a case of deception or whether the presence of another male merely changes the original male's motivation, but it represents an audience effect, which is in turn consistent with awareness on the part of these fishes. Audience effects have been demonstrated in several other fish species and can vary according to factors both external (e.g., sex of audience) and internal (e.g., their own reproductive state).[20] Such behaviors reflect the ability to scrutinize, compare, remember, recognize individuals, make subtle discriminations, and decide.

Lacking the sort of linguistic minds we have, animals may be more attuned to details we might disregard. Biologists who study ravens have to disguise themselves to avoid being harassed and avoided when they try to recapture birds. One bad experience and the ravens will know to avoid specific humans in the future. Stacia Backensto, who studies ravens in Alaska, dons a beard, a wig, and duct-taped pillows around her midriff, then assumes a different walk. It takes her thirty minutes of prep time to become a frumpy old oil field worker.[21] Noted raven observer Bernd Heinrich has resorted to wearing a kimono. When David Mech changed clothes, wild wolves he'd been studying for weeks and who were habituated to him suddenly became shy and wary. When the weather turned colder again and Mech resumed wearing his parka and wool hat, the wolves became confidently familiar again.[22]

As part of a study at the University of Washington campus, crow expert John Marzluff and his research team began wearing latex masks of either a caveman or former Vice President Dick Cheney. The crows took no notice of either. Then one day, the research team donned caveman masks and proceeded to capture and tag seven American crows using net-guns. For at least eighteen months thereafter, these tagged birds would swoop down and scold anyone wearing a caveman mask. (They remained astonishingly indifferent toward Cheney.) Both the crows who had witnessed the netting procedure and those who had only observed caveman-averse crows scolding the cavemen showed the same pattern of antipathy toward cavemen.

Deception

Deception is widespread in animals. Once thought to be merely instinct, recent studies show it to be flexible and calculating. Deceitful acts rapidly lose effectiveness if employed too often (recall Aesop's fable of the boy who cried wolf). Also, to be effective, deception hinges on others acting honestly most of the time. At the outset, let me distinguish forms of deception that don't require any sort of mentality from those that do. It doesn't require any awareness by red admiral butterflies to deceive would-be predators into thinking they are toxic by mimicking the appearance of foul-tasting monarch butterflies. On the other hand, when a gorilla feigns having his arm stuck in the bars of his cage, then quickly removes his arm and hugs the keeper who hurries over to help, we are seeing a form of deception that is premeditated, and cognitive.[23]

Chimpanzees are acutely aware of who else is in their midst, and of the status and disposition of other individuals; this may explain their prodigious spatial memory, as we saw in the previous chapter. Not surprisingly then, they are superb deceivers. Captive chimps amuse themselves by taking water into their mouths and squirting it at unsuspecting humans. So convincing is the chimp's ability to hide their intent that even wary caretakers who know they do this can fall prey to a dowsing. The ape will stroll around the cage as if occupied with something else, only to swing around and launch the attack at the right moment.[24] When the chimp hits his target, he responds by shrieking, laughing, jumping, and sometimes falling over.[25]

Studies show that chimpanzees will manipulate the visual and auditory perception of others by concealing information from them. In trials where chimpanzees could approach a human competitor unseen and reach through either an opaque or a clear tunnel to grab food that was observable and reachable to a human competitor, all seven of the tested chimpanzees chose the opaque tunnel.[26] In separate trials where the human was facing away from the food but one of two clear tunnels made a loud noise when opened, the chimps chose the silent tunnel. In a similar study, chimpanzees competed with a human for access to a contested food item. Eight individually tested chimpanzees chose to approach the food via a route hidden from the human's view, even though it was sometimes the more circuitous.[27]

Monkeys also know the value of keeping quiet in the presence of a competitor. Researchers at Yale University presented twenty-seven wild and free-ranging monkeys with two visually identical containers of food; one rattled when handled, and the other was silent. If a human sitting nearby faced the apparatus, the monkeys showed no preference for either container. But when the human's gaze was averted, the monkeys showed a strong preference for the silent container. Thus, the monkeys attempt to obtain food silently only in conditions in which silence is relevant to obtaining the food without risk of detection by a bystander.[28]

Keeping up appearances is something we share with other apes. When the chimpanzee Yeroen had been injured by his alpha comrade Nikkie, he adopted an exaggerated limp whenever he was in sight of his attacker. Male chimps often form power alliances, and Nikkie relied on Yeroen's support to maintain his position in the group. By limping, Yeroen exploited the situation to encourage sympathy and to remind Nikkie that it is never a good idea to do harm to an important ally.[29] This is more than mere deception; it is long-term strategic scheming.

Some birds use a form of acoustic deception. Shrikes are medium-size birds that prey on large insects and other small animals such as mice and lizards. Shrikes often perch with mixed flocks of birds, and benefit by catching desirable insects flushed by the smaller birds' activities. Flock members benefit by the increased vigilance of other birds, and also by alarm calls given by alert shrikes in the presence of a predator, such as a hawk. Occasionally, a shrike will utter an alarm call when there is no predator about, particularly when one of the smaller birds has captured a desirable insect. The sudden danger signal may cause the frightened bird to drop the insect, which the shrike then snatches.[30]

There are many other examples of deceptive alarm calls. Great tits use them to scatter competing sparrows from a feeder; vervet monkeys fake them to help settle boundary disputes with neighboring vervet troops, causing them to flee. Lower-ranking spotted hyenas will give an alarm cry during a feeding frenzy, causing others to scatter and freeing up space at the carcass. Increasingly known for their ingenuity, hyenas have also used this ploy to spare another from bullying.[31] This is evidence that these signals are not merely a stereotypical stimulus-response action.

Vervet monkeys may also sometimes withhold making an alarm call even though the situation warrants one. When they presented predator stimuli to vervets in Amboseli National Park, Kenya, Dorothy Cheney and Robert Seyfarth found that adult females alarm-called significantly more often when with their offspring than with unrelated juveniles. Similarly, adult males alarm-called at higher rates in the presence of adult females than in the presence of other adult males—presumably because doing so earns them credit for being more vigilant and protective. This sort of deception may be particularly effective because it is hard to detect a cheater who withholds making a signal.[32]

Scientists had thought the aerial alarm call given by roosters was just a reflex response to the appearance of a hawk flying overhead. But experiments in which computer-generated images of hawks are flown over have shown that the call is under voluntary control. A solitary rooster who sees the image remains silent even though he may take flight; but if a hen is present, he utters the aerial alarm call. Chris Evans, an ethologist at Macquarie University in Australia who specializes in animal communication and social behavior, describes this call as a "conspiratorial whisper," which helps to reduce the rooster's risk of being detected by the hawk while simultaneously warning the hen.[33] It turns out that alarm calls—and not food-solicitation calls—were the strongest predictor of mating success in roosters.

Deception can mean the difference between life or death. Playing dead is a form of deception reserved for such situations. Prey who have been attacked but not killed by a predator may feign death in a last-ditch attempt to make an escape. Studies show that death-feigning prey will open their eyes gradually every thirty seconds or so to survey the predator situation. If the pretender still sees two eyes—typical of most predators looking—then he slowly recloses his eyes. If the coast appears clear, the prey is more likely to make a dash for it. If you've ever pretended to be asleep in another's presence, perhaps you have opened your eyes a shade to see if you're being watched. The stakes are lower, but the ploy is the same.

A scene captured on film in the BBC's production *Big Cat Week* shows apparent death-feigning by a vulture. A lioness ambushed a kill surrounded by vultures, swatting one of the birds to the ground with two whacks. Lions aren't especially partial to vulture meat, and she took little interest after a preliminary sniff. Her cub arrived, further distracting her, and as the two began to amble off, the vulture suddenly sprang to life and flew off. Had the large bird continued to struggle, he likely would have come off worse.

Researchers Kurt Kotrschal and Thomas Bugnyar conducted a study with ravens that showed not only their capacity to deceive but also their ability to adapt to the deception. Four captive ravens had to search and compete for food hidden in color-marked, artificial caches. An enterprising but subordinate male raven, who found most of the food, was soon being displaced by a dominant raven who learned to follow him. Learning that going straight to the bonanza was an invitation for displacement, the subordinate took to going first to unproductive food stations, giving himself a head-start to the feast once the dominant's attention was focused on opening the box. Soon, however, the dominant caught on to the ruse, and he stopped following the subordinate's lead, instead inspecting caches for himself.[34] Ravens will also move behind visual barriers to obstruct the view of potential observers; in turn, observers watch from a distance and position themselves to be minimally conspicuous to the caching bird. Another type of deception practiced by ravens is to make false caches by burying a stone or some other inedible item, or nothing at all.[35] Ravens are also known to use different tactics for raiding the caches of wolves than those they use for taking from other ravens.[36]

When a herd of females comes onto the territory of a male impala (an elegant, mid-size African antelope), he will sometimes use an amusing form of deception to try to encourage them to stay—or rather, to not leave. Suddenly stiffening in an alarm posture, he focuses his gaze on some distant object as if he has just spotted some danger over in the adjoining territory. The ploy is to make the females reluctant to wander from the perceived relative safety of the male's territory. Biologist Cynthia Moss, who describes this behavior, is careful to use quotation marks around the word "pretend" so as not to assume that the male impala is necessarily aware of his ploy.[37]

Gray squirrels practice food-burying deception. Close observations have found that these rodents will—in addition to burying nuts—dig and cover empty holes.[38] Not surprisingly, this "deceptive caching" occurred more often in the proximity of other squirrels, and it was found to be effective in reducing the likelihood of theft by "surrogate cache pilferers," the humans studying them. Crows, jays, and other squirrels are all known to "cache in" on another squirrel's buried larders. Once a squirrel has been purloined, s/he is more likely to engage in deceptive caching, as well as to bury nuts in places harder to reach, such as in trees, under bushes, or in mud.

In sum, animal deceptions involve flexibility, restraint, context sensitivity, and—because there is a cost to using it too often—rationing. Each of these features invokes an understanding of how to use it to deceive effectively.[39]

Rising Rodents

Like most of us, I have borne our usual prejudices toward rodents. On the surface, it may seem that there is not much about these "lower mammals" to admire, with their ordinary bodies and apparently humble lifestyles lived mainly out of our view. But there's more to rodents than we may realize. Not only are they fabulously diverse and successful—they number around 1,200 species, and there is estimated to be about one Norway rat for every human—they are also clever and complex, as the many rodent examples used in this book serve to illustrate.

There is perhaps no better champion of rodenthood than the beaver. The world's second largest rodent, both European and North American species have survived human onslaughts that nearly made them extinct. They remain scarce in Europe, but North American populations have fared better; historically they are thought to have numbered about sixty million and currently are estimated to number ten to fifteen million.

Beavers must tackle various problems and solutions arising from their aquatic existence, and they have a remarkable grasp of how water works. Let me summarize just a few of their many accomplishments.

Beavers' use of sticks and other vegetation goes far beyond instinct. In particular, human efforts to thwart beavers' attempts to quell the flow of water have yielded planning and flexibility by the beavers. When beaver expert P. B. Richard inserted long drainpipes through beaver dams to allow water to flow through, the beavers first stuck mud in the dam itself near the drainpipe, near the sound of the running water.

When that failed to stem the flow, they soon found the pipe opening upstream and plugged that. When another pipe was placed near the bottom, the beavers devised a slightly different solution: they built up a platform from the bottom until it plugged the pipe. When researchers kept beavers in an artificial pond drained by a pipe with 8 mm holes in its protective cap, the inmates took to fashioning pencil-shaped sticks and plugging the holes with the tapered ends.[40]

When captive beavers build a lodge, they cut and peel sticks of two different lengths, the longer of which they use for the roof and the shorter for the entrance. Beavers have also been observed gathering material they need before starting to build, which shows forethought.[41]

Beavers use various techniques to ease the transport of logs from forest to pond. Sometimes they will install ramps over immovable obstacles. They will also dig long canals into the forest so they can more easily float (rather than drag) large logs and branches to their pond. In midwinter, beavers open a crack in their dams to let some water out, creating an airspace beneath the ice that allows them to venture farther from their lodges.[42] When accumulating underwater food piles or building dams, beavers will sometimes force one end of a stick into the mud or a tangle of vegetation to prevent it from floating to the surface or being carried away downstream by the current.[43]

Tool use by rodents is not limited to beavers. In March 2008 it was revealed that the degu—a midsize, highly social rodent from the highlands of Chile—can learn to use a rake to retrieve a seed that has been placed out of reach. The experiments, performed with captive degus in Japan, required the rodents to reach through a fence, grab hold of a tiny rake, and pull their favorite food (a peeled sunflower seed) close enough to reach it with their mouths. In the wild, degus adorn the openings of their burrows with sticks and stones, and they are known to be fond of playing games.

A popular press article reporting this study called it the first instance of a rodent engaging in tool use.[44] Not so. As we've seen, beavers' use of branches, twigs, and vegetation certainly qualifies as tool use. And there are other examples. Pocket gophers clutch stones in their forepaws while digging, to facilitate loosening and moving soil.[45] Naked mole rats—ultrasocial, subterranean rodents of sub-Saharan Africa—dig with their incisor teeth, which protrude from their mouths. While digging, they often place a tuber husk or wood shaving between their incisors and their lips. This appears to act as an oral barrier, lowering the risk of choking on fine particulate debris stirred up by the digging. The mole rats used these barriers only while digging in dry, dust-producing soil prone to aspiration, and not in soil that breaks into larger chunks that can't be inhaled. If these tools slip out of position, the mole rat either repositions it, seeks a replacement tool, or stops digging and leaves the area.[46]

Sentient Beings

One of the challenging aspects of writing about animal perception, cognition, emotion, and awareness is deciding what material to include from a steady stream of studies now being done. To give just a few examples of research not described here, I could have outlined studies showing that: rhesus monkeys match the number of voices they hear to the number of faces they see; grazing sheep scatter themselves across a field not at random but according to patterns of relatedness; Richardson ground squirrels gauge the urgency of a threat by assessing the number of alarm callers; and goldfish use odor cues to gauge the predation risk from a pike based on the size of prey the pike had eaten earlier relative to their own size.

Study after study is finding that animals are attuned to their living environments in subtle, sophisticated ways. Animals are experiencing so much more than we have been giving them credit for since around the time 2,300 years ago when Aristotle declared that animals lacked reason. Their experiences are often in sensory realms alien to us. We can only imagine what is like to have an umwelt that includes being able to recognize objects down to details in their surface texture using echolocation; or using the earth's magnetic field to orient ourselves during long-distance migrations; or communicating with others of their species by means of electric or heat signals.

Over thirty years of studying, living with, and thinking about animals has taught me that the nonhuman beings with whom we share the planet are no less sensitive or perceptive than we are, and that their emotionality, intelligence, and awareness make them fully worthy of our deepest concern and consideration. These creatures are not merely alive, but they have lives of their own that matter to them. When we see animals for what they are—autonomous, sentient beings with interests—we must realize that they were not put here for us. For too long, we have compared them to ourselves and found them lacking. As animal behaviorist and author Temple Grandin notes, "I hope we'll start to think more about what animals can do and less about what they can't."[47] When we transcend or cast off our egos and view them as beautifully whole, complex individuals with their own essences—we are no longer frowning before a mirror, but gazing through a window.

As individuals, animals are wonderfully complex and sensitive. As members of social groups, they are more interesting still. How animals interact with one another is the inevitable product of a planet populated by numerous life forms, each seeking a fruitful living. When an organism is responsible only to him or herself, there is no need for rules or restraints. As part of a social network, it's a different story. As highly social creatures, we soon learn the value of conducting ourselves in ways that keep things running smoothly for us, and for others. Other social species abide by rules and regulations. As we will see in Part II, coexistence requires good communication, cooperation, and sometimes selflessness.

Part II

Coexistence

Life on earth is deeply interactive. Organisms don't merely exist, they coexist. The characteristics discussed in Part I—sensitivities, emotions, intelligence, and awareness—evolved because they help animals make their livings. And for a great many animals, making a living is easier amongst others of one's kind. Interdependence lies at the root of ecology—the study of interactions of organisms and their environments. Evolution is the process by which organisms adapt to (or optimize their functioning within) their dynamic living environments. Other animals—especially other members of one's own species—comprise part of that environment, and often part of the optimization. The next three chapters focus on three broad aspects of animals' interactive lives: their communication with others, their cooperative tendencies, and how their actions benefit others. That last idea may sound controversial, but scientific interest in animal virtue is beginning to germinate.

How animals coexist has two main connections to this book. First, it reinforces our increasing body of knowledge that animals lead sentient lives. Second, social living is, for the most part, intrinsically good.

CHAPTER SIX

communicating

It is of interest to note that while some dolphins are reported to have learned English—up to fifty words used in correct context—no human being has been reported to have learned dolphinese.

—Carl Sagan

Animals use diverse means to convey information. Some of it we are tuned in to, a lot of it we are not. Like those bat calls I recorded at high speed and played back much slower, as mentioned in Chapter 1, some animals' signals occur outside our sensory bandwidths, and it is only with technology that we are able to detect and discern them. Because we are not them, not inside their minds, there is also the problem of interpreting their experiences. We are outsiders, peering in, unable to inhabit their lives. Fortunately, the more an animal is like us, the more reliable our interpretations may be. For example, studies of gesturing by other great apes reveal patterns similar to humans. Very young humans who have yet to attain a command of spoken language use similar gestures to those used by other apes, such as extending a hand to request food, raising both arms to be picked up, and using the whole hand to point. Emotions are also conveyed similarly, such as expressing frustration or anger by stamping the feet, pushing someone away, or turning away while shaking their head in protest.

But how can one know what a cuttlefish is communicating? Cuttlefish use a large repertoire of body patterns for camouflage and communication with others of their kind. Their skin—densely packed with up to two hundred pigment cells of various colors per square millimeter, controlled by a fine network of muscles, and enhanced by a layer of mirrorlike reflective tissue—is beautifully adapted for hiding and for communicating. By manipulating these pigment cells, combining colors like a painter's palette, cuttlefish can produce mesmerizing patterns and rhythms of metallic color pulsating across their bodies. They can also adjust their signals by altering the polarization of light, instantly blending into a background.

One recently discovered pattern is made only by female cuttlefish. On seeing another female, or a reflection of themselves in a mirror, females consistently, repeatedly splotched. The scientists who described the splotch display think that it might function to put other females at ease.[1] But interpreting these mysterious mollusks' intentions is challenging, to say the least.

Tools of Communication

Semaphore—a system of communicating over long distances by holding the arms or two flags in certain positions—is not a very efficient mode of communication for us. But for the Panamanian golden frog semaphore is just the ticket. These frogs live near waterfalls, where the constant din renders vocal communication useless. So while the males of other frog species woo females with beeps, chirps, and booms, these frogs have, over the course of evolution, traded in acoustic for visual signaling. When they want to get someone else's attention, they flash pale patches of skin on their limbs or the webs between their toes.

Is it communication? To find out, Erik Lindquist and Thomas Hetherington of Ohio State University presented male frogs with mirrors. The frogs signaled significantly more at their own reflections than they did at a nonreflective control surface. Staged encounters between males further showed that semaphores were not directed randomly, but instead were aimed toward target individuals.[2]

It takes light to perceive a visual signal, so, not surprisingly, all of these frogs are diurnal. Except one, which uses semaphore on moonlit nights. Of course, different modes of communication are not mutually exclusive. Both males and females of a species of frog in Borneo have been found to use foot-flagging, arm-waving, vocal sac-pumping, and open-mouth displays.[3]

Some frogs are ventriloquists, and their pure tone calls are extremely hard to locate. I once spent ten minutes fruitlessly trying to find a calling spring peeper frog. I got within inches, but whenever I turned my head it seemed the frog was calling from somewhere else. Female barking tree frogs don't have to deal with ventriloquist males, but they are presented with another location problem, and they have a neat way of solving it. Like most frogs, females tend to favor larger males as mates. Larger males may be distinguished by having louder calls, but a smaller male who is closer may sound louder. Female barking tree frogs use triangulation to locate the more promising mate. By moving about and listening for changes in the angle from which the sounds are coming, the female can detect a faster change in the direction of the smaller frog's calls. Triangulation is a more complex calculation than monitoring how the sound degrades or working out how fast calls get louder as the female approaches its source. "They're smarter than I realized," says Christopher Murphy, the behavioral ecologist who conducted the study.[4]

Recent research has detected elephants using a special acoustic technique to track others. Their feet are beautifully adapted for communicating and listening infrasonically, that is, at frequencies below human hearing. Preliminary studies reveal a high density of pressure-sensitive nerve endings at the front of the footpad and around the edges. This enables them to remain in contact with each other for weeks at a time even though they may be separated by miles of savannah. It may also act as an early-warning system for earthquakes, explaining why elephants and virtually all other large animals had already moved to higher ground when rising water from the giant tsunami pounded Asian coastlines on December 26, 2004.

Because we have virtually no close contact with whales, the vast extent of their communications with other whales remains a mystery to us. The high conductivity of water means that a whale's calls might be heard by another whale hundreds of miles away. Wops, thwops, grumbles, and squeaks are among the thirty-four different types of call so far identified in a study of humpback whale communication. Researchers monitoring humpbacks migrating along Australia's east coast recorded 660 sounds from sixty-one different groups of whales. A male's "purr" seems to indicate his amorous intentions toward a female, while high-pitched cries and screams were associated with disagreements between males jostling to escort females. The "wop" call appears to be a mother-calf contact call.[5]

Whales also show signs of acoustic culture, including a variety of dialects found in orcas and sperm whales.[6] Dialects are cultural because they are learned and not genetically coded. These dialects are characterized by different rhythms in the clicks the whales make. "Acoustic clans" is a term used to describe the clusters of dialects found among the estimated total population of 360,000 sperm whales worldwide. Other marine animals also draw information from these dialects. Unfortunately, trying to interpret whale communication is a bit like trying to interpret ancient cave paintings. We are so vastly separated from these creatures that we may, so far, only scrape the surface of what is actually being communicated.

Fish are not quiet. They produce sounds by grating jaws, clicking spines, slapping fins, and vibrating their swim bladders—air-filled sacs that control buoyancy.[7] In 2003, Ben Wilson and his colleagues at the Bamfield Marine Sciences Centre in British Columbia discovered the source of loud raspberry sounds among schools of herring: the fish were forcing a stream of bubbles from their anal region. The source of the bubbles is probably the swim bladder rather than digestive gas from the gut.[8] Members of the family of fishes that includes herrings and sardines have excellent hearing, aided by a gas-filled sac near the inner ear, which acts to amplify sound pressure. It is unclear for what purpose the herring use the "fast repetitive tick" sound (the acronym, appropriately, is FRT)—neither exposure to a fearful stimulus (shark extract) nor hunger elicited the FRTs in captive herring shoals. The leading theory is that the sounds allow the herring to locate one another and maintain contact in the darkness. Because the FRT frequencies are higher than the hearing of most predatory fishes, they can stick together without revealing their position to the enemy.[9]

There is a sobering footnote to the underwater communication systems of the whales and the herrings. Marine animal sound communication is threatened by human-caused noise pollution. Engine noise from shipping, seismic guns used for oil surveys, and naval testing exercises all interfere with the ability for these animals to hear each other.

Language Lessons

The importance we assign to human language has been one of the primary reasons why we have seen fit to treat animals according to convenience as opposed to conscience. "No process of natural selection will ever distill significant words out of the notes of birds or the cries of beasts," asserted the nineteenth-century, German-born philologist Friedrich Max Müller. Müller also declared that language requires thought and thought requires language.[10] Without a doubt, our grammatical use of language is unique and special. Even the most liberally interpreted studies of ape language do not approach what we have achieved with our spoken language.[11]

But we would be wrong if we assumed that we were the only species to have mastered a form of language. What makes language distinct from mere communication is that it includes arbitrary conventions as units of communication. If the only way I had of conveying to you that I was referring to a daffodil was to point to one, that would be communication, but not language. That I can utter a word, "daffodil," that denotes the specific object even in its physical absence makes it language. Honeybees have true language, because they also can convey discrete information with arbitrary symbols—the duration, intensity, and angle (relative to vertical) of the bees' waggle dance describes the distance, quality, and direction (relative to the sun's position) of a food source.[12] While humans may have developed language to a unique degree, we haven't cornered the market on it.

A great deal has been written about the use of sign language by great apes. Great apes—once thought incapable of using symbolic language—are not necessarily so much intellectually limited in their use of language as they are physically limited. Our great ape cousins don't have the vocal or mouth control to produce sounds the way we do. Thus, early attempts to get chimpanzees to speak were roundly unsuccessful. It was not until 1967, when Allen and Beatrice Gardner of the University of Oklahoma began teaching American Sign Language to their adopted infant chimp Washoe that chimpanzees were seen to have language competence.

Another important development in animal communication occurred that same year, when the American anthropologist Thomas Struhsaker reported the use of distinctive, referential calls by vervet monkeys in the African savanna. Struhsaker observed that these social primates use discreet alarm calls to represent the presence of an eagle, a leopard, or a python in their midst. The call for leopard would send the monkeys scampering up trees where they would perch on the outer branches out of reach of a large cat. An eagle call had the opposite effect, drawing monkeys away from the outer branches where an eagle is most likely to strike. The snake call invoked no fleeing, just vigilance of the tall grass for the presence of the slow-moving but stealthy foe. Struhsaker confirmed that the vervets understand the calls' specific meanings by playing recordings of the calls when there was no real predator present.[13]

For a while after these groundbreaking findings, it was presumed that apes and monkeys were the only nonhumans capable of using referential language. No longer.

In 1999, researchers from Macquarie University in Sydney, Australia, reported that domestic chickens used a variety of clucks, thirty different calls in all, as signals to refer to specific objects or situations in their surroundings. When, for instance, a rooster makes a distinctive "tck tck tck" call, it brings a nearby hen running to take the morsel from the gallant rooster's beak, or to search for it in the grass. When the Macquarie team broadcast a recording of the food call, hens who had recently been allowed to peck clean a floor scattered with corn kernels looked down for only a third as long as did hens who hadn't been fed (and who therefore didn't know whether or not there was any food present). This demonstrates that the call is not an automatic trigger for some reflex to search the ground; the birds respond flexibly according to their state of knowledge.[14]

This nice study is more a measure of our past reluctance to acknowledge animals' savvy awareness than it is a measure of birds' intellect. It is only because our science has recently begun to allow the once heretical notion that animals think that studies like this are being done. The chickens were demonstrating what they were doing all along and it only took someone to take notice and verify.

Teresa Cummings notices. With her husband, Terry runs Poplar Spring Animal Sanctuary in Maryland, a haven for abused and neglected farm animals. One day, as Terry and I were cleaning out one of the chicken sheds, a rooster let out a screech and a group of chooks happily scratching in the grass outside suddenly made a mad dash into their shed. As I dumbly looked at them, Terry was gazing upward at a hawk gliding past overhead. Unlike me, she'd seen this response before, and knew what caused it. A recent study documents at least two aerial predator calls in the domesticated chicken: one for large, and less dangerous, predatory birds, and one for smaller, stealthier enemies.[15]

If you think a chicken's alarm call is useful only for chickens, think again. The rabbits at Poplar Spring have also learned to dart for cover whenever they hear the hens' air-raid siren in the adjacent barn. There is abundant evidence that animals respond to the alarm signals of others. They do so because sounding the alarm is risky to the individual but it helps relatives survive. It is also a good deed that others may remember, and they may return the favor. Animals know they should take alarms seriously, and heeding these signals may be the difference between life and death.

Birds smaller than chickens have been found to have intricate alarm systems as well. "We really were surprised at just how sophisticated the alarm call system is and how sophisticated the judgment of predation risk was," said Christopher Templeton, who, as a doctoral student conducted a study of the alarm calls of chickadees, tiny songbirds that weigh less than half an ounce. Chickadees produce distinctive "seet" calls to warn of larger aerial predators, such as the great horned owls. Smaller, more agile predators, such as sharp-shinned hawks, present a greater threat; they elicit the characteristic chick-a-dee-dee-dee call. The number of "dees" a chickadee affixes to the end of her call (up to fifteen) provides specific information about the type of predator, and may also call in other birds to mob the predator. A predator who's been spotted first poses little immediate danger to these agile songsters, and mobbing tends to encourage the predators to leave.[16]

The call discrimination abilities of these small birds reveal a finely honed perception of slight differences in sounds. It's an ability at which humans excel because we must regularly discriminate very similar words, and there are many anecdotes that testify to a similar skill in birds. When Chris Chester had some corn for B, an orphaned house sparrow he had reared from a chick, he would call out "corn." Even if he was in mid-flight, B would veer back and land on Chris's shoulder to be fed. Chester tried similar words, "born, torn, morn, horn, worn," using the same intonation, none of which had any effect on B.[17] Similarly, Len Howard, a musicologist who lived for a decade among wild birds in England, observed on many occasions that great tits understand the command to "get off the bed." If she used the same intonation to reprimand them with an irrelevant command, such as "poke the fire," they wouldn't fly off.[18] Another great tit named Twist took to delivering a gentle peck to Howard's nose when she said "kiss"; Twist never did this without being asked, and he never kissed in response to other words. The bird was employing the same skills she used for distinguishing the calls of other birds, only this time in a domestic setting.

Befitting their social nature, prairie dogs have developed a sophisticated system of predator detection. Their alarm calls convey specific information about an approaching foe, including species, size, shape, and even color. When hawks or humans come into view, prairie dogs run to their burrow entrances and dive inside; if the enemy is a coyote, they watch vigilantly from the burrow entrance, or if it's a dog, they may just stand erect and watch from where they are foraging. If presented with only recordings of an alarm call in the absence of any actual predator, the rodents respond in kind, demonstrating that they understand the meanings of these different calls. The alarm calls of prairie dogs vary with geographic locale, so there are local dialects that dogs from other districts wouldn't understand.[19]

Northern Arizona University biology professor Con Slobodchikoff, who has studied prairie dogs for over thirty years, believes that they have the most sophisticated communication system of any other mammal. To date, research has shown there to be at least twenty different basic prairie dog words describing predators, with many more variations to account for modifiers, totaling about a hundred words.[20] They even have a specialized term for humans carrying guns. How, you might ask, did the researchers discover this? One member of the research team fired a shotgun into the ground near the colony once daily for five days, after he had been visible to the prairie dogs for between one and five minutes. After a two-day period during which this person did not visit the colony, his reappearance, without the gun, elicited a distinctive call unlike that for his presence prior to the gun firings, and the animals also changed their behavior by promptly disappearing into their burrows.[21] Curiously, studies in which people walked through the colony wearing different color shirts caused the rodents to modify their calls in response to the presence of each color (blue, yellow, and green). This is thought to indicate that the prairie dogs use labels to signify characteristics of individual predators.[22]

Prairie dogs have a sophisticated vocabulary of alarm calls that includes "man with gun." (Photo courtesy of Con Slobodchikoff.)

Despite such sophisticated alarm systems, prairie dog populations, once numbering as many as five billion, have plummeted by 98 percent, mainly from extermination and habitat loss. Cattle ranchers have tried to destroy them because of perceived grazing competition (generally unsupported by data). Some hunters even take pleasure in "misting" prairie dogs—shooting the animals with hollow-point bullets that cause them to literally explode in a mist of blood. The fact that there are humans who find amusement in obliterating these harmless animals makes me wonder which species is more admirable.

Probably because they are our closest relatives, chimpanzees have been an endless source of fascination, and the focus of a great deal of communication study. Consigned to a different evolutionary trajectory when they diverged from a shared ancestor with us some six million years ago, chimps never became linguistic. Instead, they have a repertoire of utterances that vary according to context and emotional state of the caller. Touch and gesture are also an important means of communication. When chimps greet, they may embrace, hold hands, kiss, or pat each other's backs.[23] Chimps have been shown to understand that a given gesture refers to a conceptual category. A chimpanzee who has learned the verb "open" as applying to a door can seamlessly transfer its meaning also to books, water faucets, and drawers.[24]

Chimpanzees generally use both visual and tactile gestures—the former when others are already paying attention, and the latter regardless of the attentive state of the recipient.[25] In a study of visual gesturing, a pair of caged chimps saw a banana that was placed out of their reach in a laboratory, and they tried to explain to the human experimenter how to retrieve it for them. When a human entered the room, each chimp held a hand out palm up in the classical begging gesture, and they pointed toward the fruit with their whole hand or an index finger. They also used gaze alternation—looking back and forth from human to banana. The apes deployed the same types of referential communication whether the banana was visible or hidden from view.[26] These findings illustrate chimps' capacity to try to influence the state of knowledge of an observer, and in this case, a member of another species. Chimpanzees clearly possess a theory of mind, for they are consciously aware of the conscious awareness of others.

Chimps will opportunistically use other communication tools made available to them. An eleven-year-old female, trained to communicate using a display board with 256 lexigrams (symbols representing words), watched as a desired object was hidden in the woods outside her outdoor enclosure. Following a delay of up to sixteen hours, a human was introduced who did not know the location of the hidden object. The chimp first did whatever it took to get the human's attention, then consistently and effectively used the lexigram display, combined with pointing and vocalizations, to direct the human to the hidden object.[27] This study demonstrates chimps' ability to recall and communicate features of an environment not present to the senses, and to show awareness that another is not aware of something.

Much of the communication among chimps, as for animals in general, is too subtle for us to notice. One prime example involved six young chimpanzees being studied in the 1970s by Emil Menzel at the Delta Primate Research Center outside Covington, Louisiana, near New Orleans. One of them (we'll call him the "leader") was introduced alone into an outdoor enclosure and shown either a hidden source of food or a stuffed snake. When this chimp was reunited with his fellows outside the enclosure, they quickly resumed their normal activities: playing, wrestling, and grooming. There was no readily apparent sign that the leader communicated his important knowledge to the other chimps. Yet, when all six were allowed into the enclosure after the leader had been shown food, the group headed straight for the food. In the "snake" condition, the chimps all entered the enclosure with the fur on their backs spiking up and approached the danger zone with extreme caution, poking at the leaf bed with sticks rather than with their hands. Either the leader chimp had conveyed information to the others, or they were superbly attuned to his intentions.[28]

Curiously, for all the value we ascribe to our ability to speak, language could have a dulling effect on the rest of our perceptions. Humans have developed and refined a communication system that can convey extremely specific information. It's possible that as a result we have become less reliant on our other senses for gleaning information from our surroundings. If I can verbalize my anxiety, what need have you to cue in to my body language, smells, or other physical or psychological signs that might accompany these feelings? If I tell you that I've just placed some fresh strawberries on the kitchen table, why should you have to detect their presence by sense of smell? In the absence of spoken language, other animals have to rely on their senses to detect sensory cues or the emotional states of others. It's one key reason why we may be less perceptive than many other animals.

I am not aware of any attempts to directly test this idea on animals, but there is some evidence from parrots that language may reduce the ability (or willingness) to solve problems in a social setting. In a cognition study, African gray parrots were tasked with retrieving a nut suspended from a string. Success required repeatedly and alternately pulling at the string with the beak and pinning it beneath the foot. The two parrots who had received no training in human language immediately solved the task. In contrast, the two language-trained parrots looked at the nuts, looked at the trainer and said, "Want nut." To the trainer's command "Go pick up nut," they both replied, "Want nut." This verbal interplay was repeated several times per trial and the parrots made no attempts to retrieve the nuts for themselves.[29] It should be added that most of the language-trained parrots' daily vocal demands are met as part of the training strategy in this captive setting. Later, when Arthur, one of the language-trained parrots, could observe other parrots and their test apparatuses, he reached out toward them and consistently repeated, "want some." It appears from this that Arthur recognizes that other parrots are also capable of understanding his command and giving him some food.

Despite such findings, language is surely a shaper of intelligence, and vice versa. The studies of David Premack, Sue Savage-Rumbaugh, and Duane Rumbaugh on the linguistic abilities of chimps and bonobos have found that the individuals who had learned to use symbols on a display board perform far better on logic tests. These animals learned to solve an analogy problem such as "lock is to key as can is to can opener," whereas those without language training failed to solve even the simplest of analogies, like "apple is to apple as banana is to ______." It is not as if the latter group are less smart, for they solve real-world problems as well as their trained counterparts. Instead, success in such tests, it seems, depends both on native intelligence and on knowing how to take such tests. It appears that human language skills just aren't powerful intellectual tools for chimps in the wild.

We err mightily when we couple the ability to speak with the ability to feel, because language is neither a source nor a product of sensory acuity. The inclusion of these examples was not meant to bolster the inflated importance of language to matters of personhood or moral integrity, but rather to provide another window onto experiences other than human.

Recognizing the Individual

When we look at another species in the wild, their members may look superficially identical to us. A fox is a fox and a cardinal is a cardinal. But it's not as if they really are identical; we just rarely get close enough to discern their distinguishing features. Dozens of gray squirrels inhabit the woods behind my home, and it is not until I stare at them a while from close range or study them through my binoculars that I notice idiosyncrasies—an ear notch, a faint rusty patch on the nape, or a darker coat, for example. It was many years before I even noticed that the males had sizeable fuzzy, wobbly parts underneath. The squirrels themselves, of course, have no trouble sorting out who is who.

When we put our minds to it, we are very good at discriminating between wild animals. Wildlife naturalist Judith Rudnai devised a system of individual recognition for wild lions that she studied in Nairobi National Park in Kenya in the 1970s. By learning the pattern of each lion's whisker spots on each side of his or her face, she was able to accurately identify lions for the rest of their lives. After only a year of studying wild hyenas, a keen-eyed graduate student can identify over a hundred individuals by their shoulders, ears, faces, or sides. Wildlife photographer Hugo van Lawick spent five years filming the chimps at the Gombe Stream site in Tanzania made famous by Jane Goodall's research. During this period, he spent much more time around chimps than humans, and he could eventually recognize each individual from a distance of a half mile or more based on differences in their physique, posture, shape, and movements.[30]

Not long ago scientists believed that chimpanzees had poor face recognition. Yet it turned out that researchers were showing chimps only human faces, presuming that they are easier to distinguish. It wasn't until chimpanzee faces were used that chimps demonstrated face recognition skills on par with ours, including the ability to recognize the marks of kinship. A study of five captive chimpanzees showed that they were able to match digitized black-and-white portraits of unfamiliar females with their male offspring significantly above chance.[31] Chimps also have good voice recognition skills. When a female was presented with photographs and recordings of the vocalizations of familiar chimpanzees, she successfully matched calls to pictures for a variety of vocalization types, including pant hoots, pant grunts, and screams.[32]

As a graduate student at the University of North Carolina, Renée Godard discovered that hooded warblers not only recognize their neighbors by voice, but that they recall them from one year to the next. This is particularly remarkable since hooded warblers migrate thousands of miles between their wintering grounds in Central America to their spring and summer haunts in North America, where the males establish breeding territories. Males quickly meet the other birds on the territory and tolerate their presence, but an unfamiliar male is considered a threat. By playing a recording of a neighboring male's song, Godard determined that these birds recognized their neighbors. Resident males reacted strongly when a neighbor's songs were played back at the opposite side of where the neighbor's territory was actually located. They had the same types of reactions when Godard presented recordings she had made the previous year. As there is no evidence that these birds migrate together or that they sing when they are in the tropics, they must be remembering the voices of their familiar neighbors at least eight months after they last heard them. Recognizing an accepted neighbor is useful because it saves time and energy for attracting a mate.[33] These experiments show not only that the warblers have good memories, but that they have expectations of who in their community should be where, and that they notice when things are out of place.

Birds can also learn to recognize individuals who are not birds. Len Howard wrote of being identified from afar, often two fields away, by birds who had been reared in her home. They didn't rely on clothing, because on one occasion Howard wore a head scarf and a blue raincoat she had purchased on her trip. Yet still several birds flew in to meet her though they avoided other humans. Some of the great tits also flew to greet Howard at the bus stop near her home, and she presumes they watched her board the bus and remembered this when the bus returned two hours later. Why would the birds have taken the trouble to remember what she looked like? Len rarely traveled without a supply of grains and seeds.[34]

Animals may switch identification modes depending on circumstances. If a zebra foal gets lost from her family group within the herd, family members frantically rush around in search of the youngster. The foal runs from one zebra to another, calling. Hans and Ute Klingel, who studied zebras for most of the 1960s, concluded that zebras can recognize one another by sight, voice, and smell. They use their unique stripe patterns by day, and it is only at night, in large concentrations of other zebras or in dense bush, that they resort to vocal recognition. Smell functions effectively only at close range. The Klingels observed that lost foals would touch noses with other zebras, but that when the foal saw its mother it would walk up to her without touching noses. Zebras who had been subjected to the disorientation of being immobilized with a tranquilizer dart also would walk in a straight line back to their family group within the herd.[35]

Instead of names, dolphins have signature whistles that function as individual identifiers. Not only do dolphins utter their own whistles, but they also mimic the whistles of other individuals as a way of getting their attention. Interestingly, wild bottlenose dolphins continue to respond to synthesized versions of their signature whistle, even after it has lost its distinctive vocal character. Thus, they appear to cue in on the phrase itself, recognizing their own whistle in the same way you recognize your name no matter who utters it. At the time, the research team concluded that dolphins are the only animals other than humans known to transmit identity information independent of the caller's voice or location.[36]

Within a year of the publication of the dolphin paper, however, researchers reported that spectacled parrotlets label family members by referring to them with distinct calls. The parent birds designate each of their chicks, and each other, with a different call. Studying the birds in their wild Colombian habitats, the research team noticed that specific individuals would return to the nest when their mother uttered a cry, despite the cacophony of other calling parrots in their midst. Analyses of these calls showed high levels of consistency in the label designating a particular individual, and clear distinctions with other individuals' labels, as we saw in the baby bats' "signatures" I described in Chapter 1.[37]

Foreign Interpreters

Animals may intend most of their own communication signals for members of their own kind, but that's not the only way they can be used. Signals contain useful information for anyone who can tune in. Many animals have learned to interpret other species' signals.

Studies from the University of Montana show that nuthatches—sprightly little woodland birds—can translate a foreign language: chickadee. When chickadee alarm calls were broadcast to nuthatches (with no chickadees present), the nuthatches reacted appropriately to the different alarm calls. "Seet" calls put the nuthatches on alert, while "chickadee-dee-dee-dee" calls sent the nuthatches flying toward the sound speaker, as if to mob it.[38]

Some birds take the use of other species alarm calls a step further—they mimic them. The drongos of Sri Lanka commonly travel and forage in mixed flocks, and they enjoy greater success with catching insect prey. To increase the size of the flock, the drongos imitate the calls of these gregarious species (e.g., orangebilled babblers, ashy-headed laughing-thrushes, and malabar trogons), drawing more to the area. The scientists were able to determine this by comparing recruitment when playing back recorded drongo calling that did and did not contain imitations.[39]

The drongos then use their mimicry abilities to engage in "false alarm calling." The false alarm calls startle other species into dropping prey, which the crafty drongo then snatches up. Drongos can mimic the sounds of eagles and hawks (aerial predators), magpies and monkeys (potential nest predators), and the alarm calls of at least two bird and one squirrel species. And they imitate the different mobbing calls other birds use for terrestrial predators (e.g., snake, mongoose) versus aerial foes. It may seem that the other birds are being manipulated. Perhaps, but they benefit, too. The drongos serve as "sentinels," giving reliable alarm signals when they detect potential predators. Thus, while other birds in the vicinity may occasionally lose an insect to a drongo, the loss is compensated by an enhanced security system.[40]

Mammals are also known to eavesdrop on other mammals' alarm signals. Two lemur species, red-fronted lemurs and Verreaux's sifakas, responded appropriately to recordings of each other's general predator and aerial predator alarm calls. Neither responded to recorded alarm calls of species that don't live in Madagascar, demonstrating that the lemurs understand the meaning of the specific calls of neighboring species.[41]

Birds have recently been discovered using the alarm calls of mammals. On the Ivory Coast, yellow-casqued hornbills sometimes fall prey to crowned eagles, but not to leopards, while Diana monkeys are vulnerable to both predators. As a result, hornbills respond much more strongly to recordings of the monkey's eagle alarm call than to the leopard call. When researchers played the eagle call through the microphone, the hornbills amplified their own calls and approached the microphone.[42] Calling in response to a predator is believed to benefit hornbills because it signals to the foe that they have been detected and are unlikely to mount a successful surprise attack. Similarly, the hornbills moved closer to the microphone because they are much less vulnerable to attack if they know where a predator is. This study was the first to report a bird responding to the alarm calls of a mammal. But, as with rabbits who take cover when they hear a chicken alarm call, I suspect these sorts of bird/mammal mutualisms are widespread in nature.

The orcas who inhabit the waters from Queen Charlotte Strait, British Columbia, to Glacier Bay, Alaska, feed mostly on fish. They do not go after seals and other warm-blooded prey. In contrast, transient orcas who occasionally pass through the area on their migrations eat seals, sea lions, porpoises, dolphins, and the occasional seabird. The resident orcas are very chatty, producing about one call every three minutes, while the transient whales produce only one call every twenty minutes. The seals know this. When the research team broadcast the familiar banter of local orcas, the seals showed no aversion; if anything they moved a little closer to the speaker. But the less frequent calls of the dangerous transients elicited a hasty retreat to shallow areas and kelp beds where seals are safe from orcas. This skill saves the seals from expending time and energy by not fleeing every time they hear orca voices.[43] It is similar to an office worker's tendency to stay put when she knows that a fire alarm is a test run.

One of the most bizarre examples of interspecies communication occurs between a lizard and an insect. Day geckos of the forests of Madagascar use a head-nodding gesture to "beg" honeydew treats from planthoppers. It works. The insects vibrate to signal the incipient treat, then they flick over small balls of honeydew—a sweet liquid they excrete after a meal of plant sap—straight onto the gecko's expectant mouth. It is not yet understood why the insect so willingly cooperates with the lizard, but one possibility is that the gecko's presence provides some protection to the insect in return for the treat, as ants do for honeydew-secreting aphids. As many day gecko species eat insects, it may also be a case of serving up dinner to avoid becoming dinner.[44]

Adjusting to the Din

Communication does not occur in a vacuum. Particularly in the acoustic realm, signals may need to be adjusted to compete with surrounding noise. Advances in technology allow us to monitor subtleties in animal signals that were beyond access in earlier times. It turns out that the beasts don't take the acoustic clutter lying down.

Until recently, bats' echolocation calls were viewed as mindless, mechanical "emissions" beyond any conscious control of the bat. But several species, such as lesser mouse-tailed bats and those Brazilian free-tailed bats we met in Chapter 1, change call frequencies when flying in the presence of others of their kind.[45] They also adjust their echolocation to avoid interference from ambient noise in their local environments. Bats foraging near chorusing insects use higher call frequencies than those foraging in silent areas. When ultrasonic insect sounds were broadcast to wild free-tailed bats, the bats' echolocation calls became higher when the pitch of the insect sounds was artificially increased.[46]

Electric fishes also adjust their signals to avoid interference with other individuals. Brown ghost knifefish, however, will deliberately adjust the frequency of discharges from their electric organ (a collection of modified muscle or nerve cells specialized for producing electric fields) to those of a rival, apparently to jam the other's signal in a display of aggression.[47]

Humans, monkeys, and birds are all known to increase the intensity of their vocalizations if there is background noise, or if the intended receiver is farther away.[48] The growing amounts of noise produced in expanding cities and suburbs is forcing animals to change their signals or move out. Birds, which rely heavily on their voices to communicate, have adopted several different responses to the din. The simplest is to do what humans do: shout louder. Nightingales in Berlin sing up to 14 decibels louder than their forest counterparts, with birds singing particularly loudly on weekday mornings when traffic noise peaks. Another tactic is to sing outside traditional morning and evening peak singing times. Robins in noisy areas of Sheffield, England, have abandoned the dawn chorus and now sing during the quieter stretches of evening.[49] A third response is to sing higher, above the urban noise frequency bandwidth. A study of great tit populations in ten European cities found that birds from all ten sang at an average of 200 hertz higher pitch than rural conspecifics (that is, rural birds of the same species). Their songs feature more original riffs, maybe to get more attention in the strained circumstances. Some birds—particularly those with low-frequency tunes such as orioles, cuckoos, great reed warblers, and house sparrows—don't have the physical ability to change their tune, and their numbers are declining in parks and gardens.[50]

The urban noise effect raises the question of what drives behavioral change. Do these little urban divas consciously (or unconsciously) adjust their singing to beat the blare? Or does natural selection simply favor birds born with higher or louder voices, or a tendency to sing later in the day? These latter, more mechanistic explanations seem unlikely given the short, only decades-long time periods at work. When we are in a noisy situation and have trouble hearing another's voice, we realize the need to raise our voices. My guess is that that's how other animals learn to "speak up," too. One way to test these competing theories would be to compare the pitch/loudness/timing of the same individual birds' songs in noisy versus quiet surroundings. In any event, proximate/ultimate phenomena work in concert: over generations, the genes of birds who consciously know to sing higher would gradually tend to proliferate relative to those singers who do not (or cannot) adjust.

CHAPTER SEVEN

getting along: sociability

"That's what you do in a herd. You look out for each other."
Manfred (the mammoth), Pixar's *Ice Age* (2002)

The most communicative animals are also the most social. Virtually all mammals and birds are social at some point because their mothers care for their young, and many will spend their entire adult lives in the company of others of their kind. For social creatures, the benefits of living among others outweigh the costs. Social living is a triumph of cooperation over competition. Living with others may mean having to share food resources, but among its many advantages are better predator detection and defense, consolidation of knowledge, a ready availability of potential mates, learning from others, and increased resistance against harsh environments.

Good to the Core

Cooperation goes to the very core of life. According to University of Arizona ecologist Judith Bronstein, "every organism on earth is probably involved in at least one and usually several mutualisms during its lifetime." Simply put, a mutualism is an interaction that is mutually beneficial. If you want to see an example of one at work, just look in a mirror. You may be surprised to know that you are a collection of cooperating organisms. This may sound odd, but mutualisms are synergistic—they involve different organisms integrating or working together to become something better than the sum of their parts.

In 1966, Lynn Margulis, a young biology faculty member at Boston University, submitted to a scientific journal a paper proposing that the cells that make up multicellular organisms are collections of cooperating organisms. Dubbed the endosymbiotic theory, the idea was not entirely new, having been articulated by Russian biologist Konstantin Mereschkowsky some sixty years earlier. But with the benefit of more advanced technology, Margulis was able to develop and defend the idea more rigorously. Nevertheless, it was a radical theory with deep implications for the origins of complex life forms. The paper, titled "The Origin of Mitosing Eukaryotic Cells," was rejected. Margulis submitted it to another journal. It was rejected again. And again. In all, Margulis recalls the paper being rejected by about fifteen journals before the *Journal of Theoretical Biology* finally published it.

Today, the endosymbiotic theory is widely accepted, and Margulis has achieved scientific stardom as much for her tenacity in forwarding the idea as for its veracity. Had she been a shrinking violet, we might still be wondering why mitochondria—cellular organelles that are largely responsible for a cell's energy supply—have their own DNA. At some relatively early point in the evolution of life, one unicellular organism engulfed another, both survived, and over the course of another few million years they evolved a mutually beneficial relationship. Thus was born the eukaryotic cell, of which all plants and animals are made. Multicellular organisms are hives of cooperation. In the words of Margulis and her son and coauthor Dorion Sagan: "Life did not take over the globe by combat, but by networking."[1] (Lynn Margulis was the first wife of astronomer Carl Sagan.)

There is another way in which you are, in every sense of the phrase, not alone. It's a humbling thought, but most of you isn't actually you. Some 90 percent of all the cells in your body are bacteria. Your gut alone contains trillions of bacteria of hundreds of different species. Despite their bad reputation, these are mostly good guys. Bacterial symbionts (organisms living in symbiosis) live in each of us in a synergistic relationship whereby they help us to assimilate foods as they feed themselves. This bacterial flora, referred by some as a *microbial organ*, has coevolved with us, manipulating and complementing our biology in mutually beneficial ways.[2]

Similarly, the growth and development of coral reefs is a mutualism success story on a grand scale. A reef's infrastructure is a megalopolis of cooperating organisms: coral polyps infused with microscopic algae, which live inside the polyp cells and photosynthesize in the sunlit shallows.[3] The Great Barrier Reef, which skirts the eastern coast of Australia, is 2,000 km (1,250 miles) in length.

The Synergy of Cooperation

Cooperation occurs when two or more individuals behave in a coordinated fashion to the gain of the participants. Nature is replete with cooperating individuals, both members of the same and of different species. Ants defend aphids, who provide them with sweet droplets of honeydew from their rear ends. Cleaner fish pluck parasites from predatory fish that could swallow them in an instant, but do not do so because a colleague is better than a cutthroat. Juvenile birds remain at the nest to help in the rearing of their parents' subsequent broods.[4]

How selfish genes translate into unselfish organisms has been a central question of modern biology. In the "survival of the fittest" view of the world, individuals are merely gene-making machines, and every behavior is ultimately in the individual's self-interest. Scientists have developed two widely accepted explanations for the cooperative and selfless behavior commonly shown by animals (humans included). The first theory, known as kin selection, was set forth in a 1964 paper by British evolutionary biologist William Hamilton.[5] By helping close genetic relatives, one is indirectly investing in one's own genes. The most prominent example is the care and protection of young. Many studies have shown that individuals will sacrifice immediate self-interests to benefit close relatives.

But there are myriad examples of animals acting compassionately toward others who are not kin—in some cases they are not even of the same species. That's where the second theory, reciprocal altruism, comes in. Developed by the American evolutionary biologist Robert Trivers in 1971, reciprocal altruism refers to the exchange of a favor now for the prospects of a returned favor later.[6] Readers should note that reciprocal altruism depends on individuals' (a) doing favors to one another, (b) remembering the favor later on, and (c) recognizing the individual who did the favor. Reciprocal altruists have minds, and feelings.

Selfish Not

It is astonishing to think that the behavior of sharing was considered unique to humans until Jane Goodall reported it in chimpanzees in the 1970s. Lifting the blinders that come from not believing in the existence of something has revealed many documented examples of animals sharing with each other. It's not as if animals magically started sharing in the past few decades. They were sharing all along. Either we were just not looking hard enough, we ignored what we saw, or our perception was biased by our belief in human superiority and uniqueness.

Once sharing behavior became credible, researchers started noticing it in species as diverse as baboons and naked mole rats, which happily share their food with colony mates.[7] Jackdaws are even more likely to share favorite foods than less preferred foods with other jackdaws.[8] We now know that it is one of the most important ways that social animals cooperate.

Capuchin monkeys have a natural propensity to share. When one captive individual is plied with pieces of fruit and another goes without in an adjoining cage, the first one will hand bits of food to the second, pushing tidbits through the mesh, if necessary.[9] When pairs of captive capuchin monkeys must work cooperatively to obtain unequal rewards, pairs that alternated which individual received the higher-value food were more than twice as successful as pairs in which one monkey dominated the higher value food.[10]

Common vampire bats are no less impressive in their sharing tendencies. In return for many nights spent in the unenviable position of lying face up at the base of hollow trees in Costa Rica, watching the social interactions of wild vampire colonies, University of Maryland biologist Jerry Wilkinson made waves with his discoveries of their blood-sharing habits. He found that females, who roost in a separate cluster from males, will often regurgitate blood to share with other females (or young) who have not fed for a day or more and might go hungry or starve. These bats form close associations with their cluster mates. In one known case, two females roosted together for more than twelve years. In darkness, vampire bats recognize individuals and treat them differentially according to relatedness (kin selection) and past interactions (reciprocal altruism). In other words, they (quite literally) nurture these friendships and alliances. In times of need, as when a mother is delivering a baby, or when a female returns from an unsuccessful night's foraging, one bat helps another, and this may sometimes make the difference between life and death.[11]

For cliff swallows, finding their flying insect food presents its own challenges. These birds, which nest colonially under cliff overhangs and on buildings, utter a high-pitched "squeak" call to inform others of the location of an insect swarm. The cost of this behavior is negligible because insect swarms are not quickly depleted, and one of the benefits is that it is easier for more swallows to successfully track the movements of the swarm, extending the time available for foraging.[12]

Rising above the Competition

When it comes to explaining behavior in animals, scientists err on the side of evolutionary pressures. They have historically also tended to focus on competitive rather than cooperative aspects of nature. Yet, as Con Slobodchikoff has noted, where ecologists have tried to find competition in action, the results have been elusive. "Sometimes ecologists have had to resort to a historical argument, suggesting that even if competition cannot be documented now, it must have occurred in the past to separate species into different niches." As Slobodchikoff explains, our culture conditions us to think in competitive ways. We compete for jobs, for money, for approval, and so on. We tend to think that for every loser there must be a winner. We transfer this outlook to our view of nature, leading to the notion that competition is nature's predominant credo.[13]

Economist Peter Corning points out that cooperation and competition comprise a false dichotomy, for cooperating animals are often in competition with one another.[14] Reproductive behavior provides many illustrations. For example, male primates often form coalitions to compete successfully against other males in the same troop. Male lions may cooperate to take over a pride, then squabble over mating rights. The males of many frog species amass in ponds and marshes, from where they call to attract females. Each male is a competitor to others, but each benefits more by being in proximity to his rivals, because females are attracted to the cacophony.

Clearly, cooperative behavior needn't be motivated by goodwill. There may be antagonistic or selfish aspects to it. Ravens often utter loud calls (called yells) at a food source, such as a carcass, that is being dominated by other ravens. As more ravens are attracted and arrive at the food, the dominant residents are unable to maintain their monopoly, creating a free-for-all. The yelling raven doesn't recruit others for her benefit, per se, but rather to improve her own prospects in securing a share of the food.[15] But whatever the underlying motives for yelling, and whatever its evolutionary basis, it remains a behavior that benefits others, and is thereby cooperative.

You Watch My Back, I'll Watch Yours

Another way that social animals cooperate to the benefit of all is through collective vigilance. Charles Darwin noted that the most common mutual service in the more socially developed animals is to warn one another of danger by means of the united senses of all. You will look long and hard to find a flock of Canada geese foraging in the grass and not see at least one bird standing sentry. The sentry is easily spotted, for he or she holds his head erect with neck straight up. I recall watching just two Canada geese grazing together near a busy road. One was periscoped while the other fed. Then, as if on some invisible cue, they switched roles in an instant.

How do birds decide on who has to stick their neck out next? Does one bird just take the initiative, or are there certain predesignated individuals who divvy up the tasks among themselves? We don't yet know the answers to these questions, but whatever the decision process, these birds are certainly capable of restraint and temporary self-sacrifice. The synergistic beauty of group foraging with alternating sentry duty is that collective vigilance is increased even while grazing individuals spend more time with their heads bent down at grass level. Lower individual vigilance with greater group size has been documented in mammals, birds, and fishes.[16]

Sentry behavior is not limited to members of the same species. Red-billed and yellow-billed oxpeckers (also known as "tick birds") spend much of their time on the hides of various African mammals, where they forage on external parasites and even take sips of blood. They return these benefits by providing an active alarm system to their hosts. When they hear or see anything alarming, especially a human being, they make chattering calls. When the host hears these calls, he stops what he's doing (including sleeping) and immediately searches out the possible danger.

Cleaners and Clients

The cooperative relationship between cleaner fishes and their clients is one of the most advanced mutualisms in nature. It involves cross-species cooperation where cleaner fishes groom other fishes—plucking parasites, algae, or other unwanted debris from the scales, mouths, and gills of their clients. The cleaner benefits by getting fed; the client by getting a spa treatment. The partners form relationships built on trust, developed and nurtured over the course of weeks or months. Clients will line up, waiting their turn to be serviced by their favorite cleaners. Business is brisk; a cleaner may have hundreds, even thousands, of interactions in a day, servicing a variety of fish species.[17]

This process requires that individual cleaners be able to recognize individual clients, and vice versa. Observational studies bear this out. In choice experiments where a cleaner could choose to swim near one of two clients, the cleaner spent significantly more time near familiar fish. Interestingly, client fish showed no such preference in these experimental trials, and the authors speculate that they need only remember the location of where the cleaner does business to achieve repeated interactions with the same individual.[18]

The best studied of the cleaner fishes is the cleaner wrasse, a small, slender reef fish with bright horizontal stripes. They occasionally cheat their clients by taking a quick nip at a fin. They distinguish between two different, overlapping client categories: predatory/nonpredatory, and resident/visiting clients.[19] Predatory clients are almost never cheated, and if a visitor is cheated he or she simply swims away. A resident, on the other hand, who has built up a relationship of trust with the cleaner, takes serious offense at the gaff, chasing the cleaner around.[20] This punishment makes cleaners more cooperative in future interactions. Cleaners also show extra consideration for clients in the early stages of their relationship. This includes applying extra tactile stimulation (gentle strokes of the client's dorsal fin with the cleaner's pelvic and pectoral fins—a fish massage).[21] Field observations indicate that prospective clients watch the performance of cleaners before deciding whether to permit a cleaner to inspect them, a behavior called "image scoring" or a "social prestige" rating.[22] Unsurprisingly, cleaners are more cooperative to a client if other eavesdropping clients are around.[23] That all this cognitive capacity resides in the brain of a fish just a few inches long should remind us that size is only one measure of intelligence, and a rather flimsy one at that.

Sensory pleasure helps to sustain adaptive behaviors, and social interactions are no exception. The relationship between cleaner fish and their clients is mediated by good feelings. I doubt that client fish think "I need to get rid of these parasites. Let's swim over to that stripy fish so he can remove them from me. I'll be healthier and more likely to be reproductively successful." Instead, they probably think something like: "There's my cleaner buddy. It feels good when he gives me those massages. I'm getting in line."

Richard Schuster of the University of Haifa in Israel is another biologist who believes in the power of feelings to motivate cooperative behavior. His research supports the rather sensible idea that animals cooperate because it feels good (emotionally) to do so, and that being in the company of others is intrinsically rewarding. When Schuster and his colleague Amir Perelberg offered rats the opportunity to earn a sugar treat by running in a maze either alone or with another rat, the rats preferred to run with company by a nearly three-to-one margin, even though the two-rat option required greater coordination to achieve the same treat.[24]

A study published in *Nature* reports that lions hunt more efficiently alone, mainly because hunting together requires sharing food, with a 90 percent lowering of food intake per lion. Yet lions continue to live and hunt together. There are obviously long-term benefits to living socially, but why would an individual give up short-term gain for uncertain future gain? Schuster believes the reason is that it feels good to

A cluster of young lions express solidarity and security mixed with curiosity as a trio of non-pride lions loiters nearby. (Photo courtesy of Fransje van Riel.)

work with others. It is encouraging to see other scientists who openly share this perspective, because the important role of pleasure in animals' lives has been largely overlooked by scientists. Today, there are over twenty academic journals dedicated to the study of pain, but none dedicated to the study of pleasure.

Helping Raise Others

For many sexually reproducing species, cooperation is woven deeply into their biology; the successful production and rearing of offspring requires that they work together. The most extreme form of cooperative breeding occurs in the eusocial insects—bees, ants, wasps, and termites—in which sterile female workers help raise their queen's brood. Cooperative breeding also occurs in many mammals, including primates, bats, rodents, and social carnivores such as wolves, lions, foxes, and meerkats. At least nineteen species of cichlid fishes in Lake Tanganyika in Africa engage in some form of cooperative breeding. In one of the best studied of these, the Princess of Burundi cichlid, there are typically about five helpers of both sexes, who may or may not be genetic relatives of the breeding pair.[25] Helpers do the lion's share of guarding the nest, and defending and maintaining the family territory.

Studies of evening bats by Jerry Wilkinson found that about one in five nursings involved females feeding unrelated pups. Because such behavior doesn't fit a conventional, selfish view of animal nature, attempts are made to try to explain it from a selfish perspective. One theory is that mothers may be getting rid of excess milk to reduce their weight prior to a foraging flight. Such "milk dumping" may also reduce risk of infection and stimulate the production of more milk.[26] I find this idea rather unsatisfying by itself, and wonder if pleasant sensations associated with nursing and nurturing might also motivate such selfless behavior. In any event, communal nursing is another example of how evolution can favor generosity and kindness, not the opposite. Nursing of another's offspring is known from at least a hundred mammalian species.[27]

Thomas Kunz, an ecologist at Boston University, and colleagues observed the birth of a healthy Rodrigues fruit bat at a semicaptive colony in Florida in 1991. During the three hours from onset to delivery and nursing, the mother bat was attended by a second, unrelated female member of the colony. The helper bat groomed the expectant mother's vaginal region before and during the pup's birth, intermittently wrapping the mother's body in her outstretched wings. In addition to grooming herself and the emerging baby, the mother also groomed the helper and twice wrapped her wings around her. Repeatedly, the mother assumed the characteristic feet-down birthing position only when the assistant was in position to help. Following birth, the helper groomed the pup, and she and a third female fanned the mother with her wings.[28]

Anecdotal accounts of birthing assistance have been recorded in African elephants, bottlenose dolphins, African wild dogs, Japanese raccoon dogs, and marmosets. It is quite common in spiny mice, for which "obstetrical assistance" was seen in two-thirds of eighty-six births.[29] Before giving birth, a female bottlenose dolphin forms a close association with another female, the so-called auntie dolphin. The auntie attends the birth and sometimes takes responsibility for bringing the newborn to the surface for its first breath.[30] Whales, too, exhibit cooperative parenting by babysitting. Mother sperm whales, who cannot take their babies with them on their long, deep foraging dives, leave their calves at the surface, to be watched over by other adults.[31]

Another variation on reproductive cooperation is when sexually mature individuals stay with their parents and help raise offspring in subsequent breeding seasons. This phenomenon is known to occur in about three hundred living species of birds, and many theories have been suggested to explain its evolutionary origins.[32] These include the kin selection benefits of helping genetic relatives, predator avoidance, gaining experience in raising a brood, improved future likelihood of inheriting a good breeding territory, and habitat saturation—the idea that helpers stay at the natal nest because there are no available breeding sites to migrate to. Whatever the reasons, when birds breed cooperatively, they are engaging in a beneficent behavior. Incidentally, it's time to add a new word to the dictionary: "grandchicks." The Seychelles warbler was recently hailed as the first example in which formerly nesting birds will help raise the young of their own offspring who have assumed command of the territory.

Male manakins—small, brightly colored birds of the tropics of the Western Hemisphere—perform song-and-dance routines to try to induce females to mate with them. Even though only one male—the designated alpha—will do any of the mating in a given season, he is assisted by one or more beta birds who act as "backup singers/dancers." The duo or trio perform their routines in an elaborate, coordinated fashion. For instance, the birds may shimmy across a horizontal twig with such rapid and tiny steps that they appear to be on a conveyor belt, or they may hop over one another like balls being juggled. We find the effect of this team effort undoubtedly more impressive to behold than if a single male were performing. Female manakins apparently share that opinion, for there is evidence for higher reproductive success among team-supported alpha males.[33] Emily DuVal, who studies lance-tailed manakins in Panama, informs me that solo males (i.e., without backups) also sometimes breed successfully, which raises the intriguing question: Why be a beta (a backup) if you can go it alone? Strictly speaking, evolutionary theory would predict that betas ultimately have greater reproductive success than do solo birds. Further research may tell, but it is an intriguing cooperative mating system.

Female house mice breed cooperatively, with two or more mothers combining their pups into a single crèche, or group. They share a variety of brood-rearing duties, including nursing another mother's pups. This sharing is not predicted by a selfish, competition-focused evolutionary system, because the extra labor of looking after another's pups would cost the helping mother more expenditure of energy that she might otherwise devote to her own offspring. Communal nursing behavior is now known to be common for many rodent species.[34]

For the young mongoose, there is an even greater level of care. Childless adults follow a "feed the nearest pup" rule, provisioning, protecting, carrying, grooming, and playing with the most vulnerable pups in the band. Escorts consistently associate with the same pup, but since it is the opportunistic pup who maintains the relationship, they are not likely to be genetically related.[35]

Mongooses' once-obscure but now popular cousins, the meerkats, are highly social and cooperative, performing several functions including huddling together to conserve warmth during the cold desert nights, hunting collectively, defending their burrows, and taking turns standing sentry duty, sharing information through a repertoire of vocalizations.[36] Another task shared by these tight-knit groups is the feeding of young. Like mongooses, meerkats also follow a "feed the nearest pup" rule. Hungrier pups beg more and are more likely to be fed if they are nearby and the morsel is large, regardless of whose offspring they are.[37]

Contrary to the cold and fierce reputation summed up in their scientific name—*Crotalus horridus*—timber rattlesnakes lead rich social lives. Pregnant females congregate in birthing rookeries to care for their young, and, being blessed with the ability to recognize close kin, spend more time with family members.[38] Rulon Clark of Cornell University, who made these discoveries, speculates that our failure to ascribe social feelings toward snakes stems from our prejudices toward their strange body form and movements, and those angry-looking, lidless eyes. Among other reptiles, active parental care is uncommon but by no means absent. Parental care has been reported in more than 140 species (about 3 percent) of reptiles.[39] While crocodilian parents don't feed their self-sufficient young, mothers (and in some species also fathers) guard the nests, assist their hatchlings' emergence from the nest pile, and gently carry or shepherd them through their otherwise perilous journey to water.

As we can see, the reproductive benefits of assistance from family members or close associates play out broadly in wild nature. The underlying causes for the evolution of such beneficence may be the restrictions imposed by habitats running at full capacity. When there is room to breed, natural selection usually favors individuals who strike out and raise their own young. But nature's limitations often favor working together, to the ultimate good of all participants.

Team Hunting

Cooperative hunting in marine life is widespread. While diving in the Red Sea, ethologist Redouan Bshary of the University of Neuchâtel in Switzerland noticed that two large, predatory reef species—a grouper and a moray eel—could often be seen swimming peacefully side by side. Later, Bshary witnessed a grouper make a distinctive head shake on encountering a moray eel. Presently, the eel swam off in tandem with the grouper. It appears that the grouper's headshake signals an invitation to work cooperatively for a meal. The signal has been seen several times since. The two fishes will consort for up to 45 minutes. It turns out that both team members have about five times' greater hunting success when they hunt together.[40] Groupers generally catch their prey in open water, whereas the slender eel pursues fishes into recesses in the reef. Trying to escape one of the hunters puts the prey at the mercy of the other. In one instance, Bshary observed a grouper first waiting for an escaped prey fish to emerge from a nook, then swim away and return with a moray eel.[41] The banded sea krait (a marine snake) and the Travali fish also hunt cooperatively on the Indonesian reefs, although it is not yet known how they signal to one another their readiness to commence.

Along the coast of Georgia, gulls, herons, and egrets assemble on the shoreline in anticipation of what appears to be scheduled teamwork—fishing with bottlenose dolphins. The dolphins corral a school of fishes, and then ambush them up against the muddy bank. The sudden surge pushes many fish ashore, where the birds and dolphins hastily snap up those they can reach. As the fishes flip about, desperately trying to escape back to the water, the phalanx of birds nabs strays missed by the dolphins. Further out to sea, sharks, dolphins, seals, gannets, and whales make a deadly hunting combination for huge shoals of sardines. As the birds dive like javelins from above, fleeing fishes are driven downward, to the benefit of deeper predators.[42]

Returning the Favor

Mutual dependence is important in many primate societies, and it is fundamental to the social dynamics of chimpanzees. Chimps are aware of their debts to others, and they practice reciprocity in diverse situations, from obvious to subtle.

One of the more subtle expressions of their reciprocity was observed by primatologist Frans de Waal, who spent several years studying the chimpanzee colony at the Arnhem Zoo. De Waal focused on how two males fought for predominance of the pack. The "power plays" of these male chimps, some of whom were introduced in Chapter 5, may sound like the plot of a soap opera, but for the chimpanzees it's serious business. After Luit deposed Yeroen as the alpha male, he cut off Yeroen's mating privileges with the colony's females. Subsequently, Yeroen supported Nikkie's bid to displace Luit. When Nikkie eventually succeeded in dethroning Luit, Yeroen immediately made his intentions plain, openly trying to mate with females under Nikkie's nose. Because Nikkie was dependent on Yeroen's support to keep his place above Luit, Nikkie had to let Yeroen have his way. This is just one of thousands of observed chimp alliances in which individuals support each other in fights and other conflicts.[43]

Chimps are judicious in choosing competent partners. Experiments show that when chimpanzees were not able to solve problems independently they would recruit a partner. And they didn't just choose the chimp nearest to them—they almost always recruited the more effective of two available collaborators based on their previous experience with them. These experiments show that apes make deliberate choices about when to ask for help and then determine the best collaborative partner—skills shared by chimpanzees and humans.[44]

Guppies do a similar thing. Pairs who engage in predator inspection show reciprocity in switching the role of leader. Captive studies in which fishes are unable to reciprocate by the sudden insertion of opaque barriers between partners show that these fishes build up trust in each other. Predator inspections are subsequently more likely and more thorough with fishes who had not suddenly vanished. The past behavior of several partners is tracked simultaneously by these account-keeping fish.[45]

Small, squirrel-size South American primates called tamarins and marmosets are strongly interdependent on each other for survival and the successful rearing of their babies. Their relationships are close, trusting and secure; aggression is very rare. When presented with a problem that requires cooperation to solve—a tray of food that can only be dragged toward two caged individuals if both pull on their end of a string—cotton-top tamarins usually pull whether or not the food would be delivered to them or to their partner. In tests with marmosets (close cousins to tamarins), one, the donor, was given a choice of pulling or not pulling a tray to within reach of another marmoset. The donor pulled the tray, on average, 20 percent more often when there was another monkey present than if there was no other monkey. Furthermore, donors were equally generous to unrelated recipients than to related ones, so this behavior appears to involve more than merely favoring kin.[46]

Rats are no less giving. A recent experiment conducted with rats at the University of Berne, Switzerland, demonstrates a form of cooperation never before documented in nonhumans. Claudia Rutte and Michael Taborsky show that female rats are more likely to help another rat if they themselves have previously received help, irrespective of the identity of the partner. Rats who had been trained to pull a stick in order to produce food for a partner pulled more often for an unknown partner after they had been helped than if they had no experience of receiving help. Rats showed a 20 percent increase in propensity to help after receiving anonymous help. The rats pulled the stick only rarely when they were alone in the cage. Using a study design that allowed Rutte and Taborsky to fairly confidently rule out conditioning, social learning, and imitation, the researchers gave the rats' beneficence the rather comical term "anonymous generous tit-for-tat."[47]

When one animal helps another, scientists set to work on finding an adaptive, Darwinian explanation for it. While it may be true that the chimp who shares food with another is bolstering his or her status in the community or securing reciprocal kindness in future, this evolutionary context does not preclude the possibility of the chimp's also feeling good by doing good. We know this from our own experience, as subjects of nature's system of rewards and punishments. It is no coincidence that kindness feels good, for it benefits us. When we acknowledge the moral basis of our own behavior, it is hypocritical to then deny any moral foundations to the behavior of other animals.

Solidarity

Solidarity is an exceptionally successful strategy in nature. Banding together with others allows the members of many species to overcome adversities and challenges that individuals acting alone cannot.

Hymenopterans—ants, bees, and wasps—are the most social of all animals. Their dedication to the common cause is exceeded by none, humans included. And they are very successful. These champions of solidarity make up only about two percent of all insect species, but their worldwide biomass amounts to around 80 percent of all insects. On a visit to Mexico in 2008, I discovered a colony of leafcutter ants transporting vegetation from a pair of trees to their nest. Leafcutting ants, who discovered agriculture about 50 million years before humans did, don't eat their harvest directly—they grow fungus on it and they eat the fungus. At all hours of the day, ten to twenty thousand ants could be seen scurrying in a foot-wide column. Most had a hunk of leaf sticking up from their jaws like a sail. Leafcutter colonies have as many as five different castes, and I noticed that a few of the smallest ants clung to the green sails carried by their larger kin. In their book *The Superorganism: The Beauty, Elegance, and Strangeness of Insect Societies*, ant authorities Bert Hölldobler and Edward O. Wilson speculate that these hitchhikers protect the carrier ants from parasitic flies that try to steal into the nest and pilfer from the ants' fungal gardens.[48]

One of the core benefits of solidarity is safety. We've already seen how collective vigilance is improved when there are more eyes (and other senses) on the alert. Baboons who cooperate to defend the troop against a predator are more likely to thrive as a troop—with all the group's various benefits—than is a troop without cooperation.

Naturalist and wildlife photographer Hugo van Lawick gave a detailed account of a group of plains (or Burchell's) zebras defending themselves against an attack by African hunting dogs. The dogs targeted a foal, but the rest of the zebra herd slowed their running speed so the foal and his mother would not fall behind. The dogs persisted and eventually managed to separate the mother, foal, and a yearling from the herd, which disappeared beyond a ridge. The dogs made several lunges at the foal, but each time, the mother and the yearling managed to repel them, even when the dogs tried to grab hold of the mother's lip, the pain of which usually immobilizes a zebra and makes a kill more likely. Just as the dogs' efforts became more frantic and the prospects for the foal appeared especially grim, the ground began vibrating. Ten zebras reappeared over the brow, running toward the confrontation. They closed ranks around the mother and her two offspring, and the dogs were soon overwhelmed. When the herd then galloped away; the dogs pursued for only about 50 yards before giving up.[49]

Zebras also work cooperatively to aid old and slow members in their midst. During their zebra studies (mentioned in Chapter 6), Hans and Ute Klingel time and again "saw wounded, sick, or old animals being taken into consideration by the other members of the group," who waited and slowed their marching pace.[50] The speed of the entire group would be reduced to that of their slowest member. The Klingels never witnessed any signs of antagonism toward the feebler animal.

An amateur video filmed by an American tourist around 2006 bore witness to a dramatic scene at a lake in Kruger National Park in South Africa. A group of lions hunkered down on a small ridge as a large herd of Cape buffalo approached. Two adult buffalo and a calf were well in front of the herd, and they got to within 15 yards of the lions before noticing them. They turned and fled. The lions pursued. The calf was tackled into the water while the rest of the herd galloped off. It looked dire for the calf. Then it got worse: as the lions latched onto the calf (still standing) and began to pull him onto the bank, a large crocodile clamped onto the calf's hind leg and tried to pull him deeper into the water.

If there are such things as hopeless situations, having one end in the clutches of several lions and the other in the jaws of a crocodile must be high on the list. The lions won the tug-of-war and pulled the calf a few meters up onto the bank. Another minute passed. Then the scene took another dramatic turn: the huge, brawny mass of some two dozen buffalo returned, quickly approached, and began to menace the lions, who became increasingly distracted from their quarry. One lion was caught off guard and tossed by a buffalo's horns several feet into the air. As the unnerved lions began breaking off from the pride, the calf miraculously rose to his feet and walked back into the herd. The buffalo continued to harass the remaining lions until the last one had been chased off.

However you look at it, whether on land or sea, getting along with others is an enormously successful life strategy. As we're seeing, social living fosters kindness toward others. It is not a big step from social living to virtuous behavior. Animal virtue is one of the frontiers of animal behavior study. Let's see what we've discovered so far.

CHAPTER EIGHT

being nice: virtue

The more I give to thee the more I have.
—William Shakespeare, *Romeo and Juliet*

In *A Primate's Memoir,* Stanford University primatologist Robert Sapolsky describes a confrontation he witnessed between a lion and three baboons. Two of the baboons were youngsters who had wandered a bit too far away from the adults, and on seeing the approaching lion, had fled to a sapling. Too scrawny to support their weight, the sapling bent horizontally and the pair of youngsters looked like ripe fruit ready for the lion's plucking. Benjamin, an adult male member of the baboon troop, saw what was happening and hurried over, bravely putting himself between the trapped pair and the predator. Though he could easily have fled to a nearby tree, Benjamin menaced and lunged at the lion, who approached to within five feet, then turned and walked off. Sapolsky is fairly certain that Benjamin was not the youngsters' father, though even if he had been, his actions would still be admirable.[1]

Virtue is moral excellence. Although it is impossible to confirm whether Benjamin was motivated by a desire to do what is right, there can be no question that his behavior was virtuous. He saved the lives of two terrified children, and at great personal risk.

Historically, scientists have sought to explain away meritorious animal behavior as merely adaptive. Our reluctance to ascribe goodness to other animals is borne of old prejudices. Raised on notions that we are alone at the top, the only ones bearing souls, and that all other creatures are savages, we're hardly encouraged to consider that animals might be virtuous. As with so many of our earlier views, that's changing. The following pages examine exciting new evidence that humans are not the only ones with a sense of right and wrong, an awareness of fairness, or just plain consideration for others.

Avoiding Conflict

In 1990, I played a game of ESS Croquet on the lawn of Susan Reichert, a biology professor at the University of Tennessee. ESS stands for Evolutionarily Stable Strategy, a concept formulated in the early 1970s by renowned British evolutionary biologist John Maynard Smith. Among the dozen or so players that day was Maynard Smith himself, who was visiting from the University of Sussex. Each of us was given a different "survival strategy" for the game. The Hawk would ignore the hoops and attack any player within range. Dove was equally committed to fleeing anyone who got near. Grudger doggedly continued to retaliate against anyone who had done him wrong. By chance, I was assigned Tit-for-tat, which meant that I was to treat others as they treated me, returning a favor or an ill deed just once before getting back to the business of knocking my ball through the next hoop.

I have to confess I don't recall who won that game, only that Maynard Smith and I were the first to finish. Though I am not skilled at croquet, my own success on the playing field that day was probably not a chance event. In computer models that pit survival strategies against each other, Tit-for-tat has been shown to be a winning strategy.

Tit-for-tat is an ESS because, if practiced by nearly all members of a population, it cannot be supplanted by any other strategy. Individuals who treat others in kind tend, over time, to flourish while more aggressive strategies falter. It's a theoretical illustration of the fact that aggression and brutality are not favored by nature.

When Maynard Smith and his coauthor George Price introduced the ESS concept in a seminal paper in 1973, they drew inspiration from the relatively peaceful means by which animals settle conflicts.[2] When agoutis fight, the loser signals defeat by lying on his side, with eyes closed and legs stretched out. Like the defeated wolf who exposes his vulnerable throat and belly to his stronger rival, or the gull who points his closed bill skyward, the rodent communicates his accepted social position, and nobody gets hurt. Male guinea pigs squeal with fury during aggressive encounters with other males. In defeat, however, the male signals his loss by presenting his rump and protruding his most vulnerable part: his brightly colored scrotum. In human society such a gesture might be construed as the height of insult, but it is a gesture of defeat—the guinea pig's way of raising the white flag. And, like most such gestures in nature, it has a mollifying effect on the victor, who accepts and commits no further violence.[3]

Being civil is a prerequisite for living in groups. To appreciate the unwritten moral codes of conduct of group-living animals, it helps to consider what could happen if there were no such code—if individuals behaved completely selfishly. Large lions would kill and eat small ones. Predators could get an easy meal by turning on anyone who was asleep. Such behavior would soon lead to group disintegration.

A herd of walruses hauled up on a rocky outcrop in the Arctic illustrates how social animals get along without inflicting grievous harm on one another. These animals are enormous. The largest adult males can weigh two tons, and females commonly exceed one ton. Protruding from their maws are two long, daggerlike tusks. To see these giant blobs of blubbery flesh crowded tightly on their beach is to wonder that they aren't constantly jabbing and stabbing one another, and it is an achievement of care over clumsiness that no tusk is bloodied at the cost of another. Though most people think of walruses as nothing more than a lumbering hulk with tusks, they are actually to our knowledge the most cognitively and socially sophisticated of all pinnipeds (seals, sea lions, walrus). Strongly gregarious, they are smart, playful, creative, and, above all, civil. Walruses will share food, come to another's aid, and even nurse another's baby. According to science journalist Natalie Angier, a captive walrus will greet you by pressing her head into your stomach and pushing. If you want to avoid being pushed over, it's best to place your palm on the walrus's snout and push back. If you want to be friends for life, blow in the walrus's face. They will lean into your breath, bleating, grunting, and snorting for more.[4]

I am not claiming an absence of violent conflict in animal societies. Dangerous fights definitely occur. If two bull elephants are very closely matched in size, they may get in a battle of strength, with trunks and tusks locked. Two elk bulls may clash antlers for ten minutes or more until one tires or loses his nerve. Deaths are said to occur, but these are extremely rare events; in the great majority of cases, two animals will avoid violent conflicts at the outset with displays of power or gestures of appeasement. The characteristic territorial confrontation between a pair of male wildebeest involves no physical contact. Both animals drop to their knees, and sometimes thrust their horns violently into the ground, or pluck grass in what looks like redirected aggression. A retreating male will often adopt a "grazing attitude," which appears to appease the aggressor.[5]

Randall Lockwood, an ethologist now with the American Society for the Prevention of Cruelty to Animals (ASPCA) who studied wolves in the Alaskan wilds and at the St. Louis Zoo, has seen hundreds of wolf spats. Only one ended in bloodshed, and that was when a wolf bit his own tongue: "They use no more force than is necessary to get what they want out of a situation. . . . The wolf pack is based on affection, appeasement, and solicitation."[6]

One of the most powerfully pacifying effects of group living is the importance of reputation. Others know who you are, and they keep tabs. Sometimes a male baboon who is being attacked will pick up a young infant and hold it to his belly as a mother would do. While it may seem a rather devious and ignoble tactic, it usually inhibits the attack, because the other baboon doesn't want to appear to threaten or risk harming the infant.[7]

Chimpanzees practice proactive peacemaking. Grooming is known to lower stress hormones and relieve tensions in primates (humans included). In a captive colony of nineteen chimpanzees, adults groomed more often in the period just before regular scheduled feeding times. Conflicts are most common before and during feeding times, and grooming is known to benefit participants by reducing tension and fostering good feelings among members of the group. Thus, grooming before feeding appears to be a conflict management strategy to increase levels of tolerance around food. Another study documenting grooming as a stress-reducer noted that the effect was more marked when food was clumped and therefore more likely to cause disputes than when it was dispersed.[8]

If strife does arise, there are other peacemaking options: reconciliation and consolation. Approaching and hugging one another is one way apes and monkeys may reconcile following a conflict. For example, when a primate is wounded in a fight, the aggressor will often approach the victim afterward to inspect and tend to the wound. Apes will hover around the victim of an aggressor's attack, offering consolation with hugs and kisses.[9] And it works. Consoled chimps engage in less stress behaviors such as scratching or self-grooming.[10] In baboon society, following a threat or an attack, the aggressor often gives a distinctive grunt, which has an immediate and striking effect on the opponent: the victim stops cowering and grimacing and relaxes. While there are very few reconciliation studies on animals other than primates, there is nevertheless evidence for it in gray squirrels, spotted hyena, lion, dwarf mongoose, bottlenose dolphin, feral sheep, domestic goats, and domestic cats.[11]

Fairness Awareness

There is emerging evidence that some animals have a sense of right and wrong. They share with humans the neurological and chemical features of moral nature.[12] These features include the role of the brain's amygdala and hypothalamus in feelings like empathy, and the actions of the neurotransmitters dopamine, serotonin, and oxytocin, which mediate these feelings.

One way to test moral awareness in animals is to see how they respond to unfairness—that is, to situations in which individuals receive unequal rewards for the same cost. For instance, a chimpanzee is more likely to refuse food for a token if she can observe another chimpanzee receiving an appetizing grape when she is only being offered a slice of cucumber. Furthermore, chimps are more likely to take exception if they are less familiar with the chimp who is getting the grapes. But if the other chimp is part of the subject's tightly knit social structure—which in chimps is characterized by intense integration and social reciprocity—inequity is usually tolerated.[13] Humans are also more tolerant of inequity in closer relationships.[14]

When a similar study was done with captive capuchin monkeys, they too were more likely to reject a cucumber slice after seeing another capuchin receive a more attractive grape. The unfortunate monkey was persistently much less likely to exchange the token for the cucumber. These observations suggest that the monkeys recognize inequality, and are willing to give up some material payoff in hopes of getting a more equitable outcome. The lucky grape recipients ended the experiments in a cheerful mood, while their partners typically sat sulking in a corner. The possibility that the monkeys might have been reacting merely to the presence of the higher-value food is lessened by the results of a control experiment in which grapes were visible but were not given to either monkey; in that case, the animals' reaction decreased markedly over successive trials.[15]

We need not stop here in our search for good-natured animals. There may be at least as much virtuous behavior in social carnivores as in primates. Wolves, meerkats, and wild dogs have a social organization—divisions of labor, food sharing, care of young, dominance hierarchies—that resemble those of our own ancestors.[16]

In a study of fairness in nonprimates, a research team led by Friedericke Range placed pairs of dogs side by side in front of a person. In plain view of both dogs was a bowl of treats (sausage and dark bread). Each animal was asked in turn to offer his or her paw to be shaken by the person. The researchers recorded the number of times each dog offered a paw under various conditions.

The results were unequivocal. When both dogs were regularly plied with treats, both gave their paws for nearly every trial. When neither dog received treats, the dogs only shook paws in about twenty of thirty trials and required more verbal prompting. Most interestingly, when only one dog received treats in return for a paw-shake, the other dog declined the handshake sooner, only offering a paw an average twelve times out of thirty, and acting decidedly more agitated in the process. In control trials with only one dog present, the paw was offered significantly longer when no treats were given, ruling out fatigue or frustration to explain the refusals in the two-dog setting.[17]

Can birds have a sense of fairness? A phenomenon that has been observed for centuries on China's Li River suggests that cormorants might have such a sense. The cormorants here are used for fishing by the locals, who bind them to the fishing boats by long tethers. When the birds dive for fish, a noose round the neck prevents them from swallowing their quarry. To sustain their slaves, the fishermen have always loosened the noose after every seventh catch, allowing the bird a reward for services rendered. On those occasions when, either by neglect or indifference, a fisherman fails to loosen the noose, the birds refuse to dive. Not only does their refusal indicate a sense of fairness, it also indicates an ability to count to seven.[18]

Starting Early

Just as the social nature and moral development of animals has been underestimated, so too did leading theorists of the early twentieth century underestimate human children. Freud characterized children as motivated by either selfish or aggressive impulses, whereas John B. Watson and B. F. Skinner viewed them as asocial, each a blank slate on which later experience would begin to etch its effect. A reflection of this thinking is found in William Golding's famous novel *Lord of the Flies* (1954), in which a group of schoolchildren turn to savagery after becoming marooned on a tropical island.

Today, it is understood that children's moral development begins very early in their lives. Acts of sharing, caring, helping, and altruism are commonplace among toddlers. At just one year of age, most children show comfort responses toward others, a behavior that becomes more nuanced in two-year-olds. They are even capable of sympathy and forgiveness.[19]

Like human children, chimpanzees also show spontaneous goodness from a young age. When three chimps aged between three and four and a half years were given opportunities to help humans who had dropped or misplaced nonfood items (e.g., marker, clothespin, book), all three chimps helped by reaching for the object that had fallen out of the human's reach. On no occasion did the chimps receive any benefit, such as a reward or praise, for helping.[20]

The tendency to help carries into the adult lives of chimps. In experiments at the Max Planck Institute, twelve of eighteen semi-wild adult chimps went out of their way to help an unfamiliar human who was struggling to perform a task (reach a stick), even when it required having to climb over an eight-foot rope barrier for no reward. Human toddlers will do the same thing. Chimps who had been taught how to unlatch a door also helped unfamiliar chimps struggling to get through the door—by unlatching the door for them.[21]

In a related study, an adult chimpanzee reared in captivity was shown videotaped scenes of a human actor struggling with one of eight problems, then shown two photographs, one depicting an action or object (or both) representing a solution to the problem. On seven of the eight problems, the chimp consistently chose the correct photograph. For example, when the video depicted a human shivering violently while standing next to a disconnected portable heater, the chimp selected a picture of a connected heater.[22]

I find these studies all the more revealing in that they involve chimps helping a member of another species, a species quite close to them.

Holding Back

Austrian writer Karl Kraus (1874–1936) observed, "How powerful social mores are! Only a spider's web lies across the volcano, yet it refrains from erupting." If you've ever been in a public place, fuming mad about something but keeping a calm exterior, you know just what he's talking about. Restraint is an essential ingredient of civilized behavior. It takes discipline to resist the desire for an immediately available reward. Oscar Wilde quipped, "I can resist everything except temptation." Animals resist temptation in their daily lives, just as we do. Social living demands it.

Once again, our close cousins, the great apes, are a shining example of restraint. Four chimpanzees and an orangutan showed more self-control than many human children in an experiment where up to twenty chocolate pieces were placed one at a time into a bowl. An ape could consume the treats at any time during the placement, but provisioning was immediately curtailed at that point. The apes quickly learned to wait until all twenty chocolate pieces had been placed in the bowl.[23] In other experiments where rewards were given for patience, humans caved quicker than chimps. If getting a bigger share required waiting two minutes, 72 percent of chimps waited for the duration, compared to only 19 percent of human subjects. (This experiment might have sold the human children a bit short; studies done in the 1960s found that about 30 percent of four-year-old children were able to wait for 15 minutes without eating marshmallows, cookies, or pretzel sticks in return for a double portion.)

Can a house sparrow resist temptation? Chris Chester's tame house sparrow, B, had to learn not to peck or tug at moles and other spots on Chester's skin, but B never directed a peck at anyone's eye, despite their often being littered with bits of "sleep goo" first thing in the morning. However, B had not forgotten that he had an effective weapon. If B was frustrated, wanting Chris to play with him by tossing a cap for him, for example, B would occasionally nip Chris with enough pressure to be painful.[24]

Sea lions are also masters of self-control. Four captive male sea lions learned significantly more quickly to opt for the smaller of two offerings of food in return for a larger payoff than did orangutans or rhesus macaques in other studies. Differences of this sort might relate to the social structure of the species in question. Holding back is more costly for a macaque because, living in large, hierarchical groups, a macaque is more likely to have something snatched away than is the comparatively solitary orangutan. Sea lions, while gregarious on land, swim alone during daily hunting and do not have to compete directly with other sea lions for access to fish.[25]

Rats have been shown to restrain themselves when they know their actions would cause pain to another individual. A 1959 study with the

Sociable weavers and redheaded finches in the Kalahari Desert take turns drinking from a mug of water set out by the photographer. (Photo courtesy of Fransje van Riel.)

title "Emotional Reactions of Rats to the Pain of Others" described rats who would stop pressing a bar to obtain food if doing so delivered an electric shock to a rat next to them.[26] Another experiment confronted rats with a fellow rat who was strapped tightly into a suspended harness. By pressing a lever, the witnessing rat could lower the other to the floor. This is what witnessing rats did.

Another form of restraint is self-handicapping. Many animals self-handicap in play situations. In Arabian babblers (a kind of passerine bird), it is invariably the more senior of two playing individuals who does this, usually by rolling onto her back, where the belly is exposed and she cannot run away or charge without first rolling back onto her feet. Cynthia Moss described to me the way an older elephant will self-handicap during play with a smaller one: "What is touching about these play bouts between unevenly matched individuals is how gentle the older one always is. He will be very careful not to scare or intimidate the younger one. I have seen big adult males lie down (like a dog) and then spar with a much younger male. Also older calves will lie down on their sides so that the small calves can climb and play on them. It shows a remarkable ability to, in a sense, put oneself in the other's place."[27]

Play signals themselves appear to also act as self-handicaps. Rolling over is a widespread play signal. When the signaler is in a vulnerable position, it conveys the appearance of a nonthreatening, friendly invitation. The bow—deployed by lions, baboons, rhesus monkeys, canids, and several birds—moves the bower's center of gravity forward, an awkward position from which to attack or flee. Larger piglets self-handicap by sitting, kneeling, or lying down when playing with smaller ones.

In most social predator groups, the top-ranking individuals eat first. African wild dogs are an exception; the yearlings are allowed to eat first. John McNutt and Lesley Boggs, authors of *Running Wild: Dispelling the Myths of the African Wild Dog,* have observed for many years this unusual "priority of access."[28]

Feeling for the Other

Empathy is the capacity to relate to how another is feeling. The early twentieth-century Russian psychologist Nadie Ladygina-Kohts raised a young male chimpanzee named Yoni. As is a young chimp's way, Yoni was often unruly, delighting in defying Nadie's authority. One of his favorite spots was the roof of her house, and Nadie's firm commands and entreaties to get him to come down were fruitless. Eventually, the psychologist discovered that the only way to get Yoni to come down was to appeal to the chimp's concern for her. By closing her eyes and pretending to weep, Yoni would leave his perch and hasten to Nadie's side to comfort her while looking around indignantly for the source of her upset. Appealing to Yoni's empathy turned out to be the only way to circumvent his defiant nature.[29]

Carolyn Zahn-Waxler, a pioneer in empathy research, demonstrated that empathy emerges before language in human children, suggesting that it is not a language-dependent trait. She also found that domesticated animals such as cats and dogs are upset by stress-faking family members, hovering over the perceived victims and putting their heads in their laps.[30] The recent discovery of mirror neurons in primates and cetaceans suggests further that empathy may be "hardwired" in these animals. With these neurons, the same part of the animal's brain is stimulated when they perform an action (say, stroking or grooming another) as when the animal sees another perform the same action. The latest species found to possess mirror neurons is the swamp sparrow, which suggests that these neurons occurred in the common ancestor of birds and mammals.

Can one mouse feel for another? When scientists injected painful irritants into the stomachs and paws of mice, then put them into a clear plastic tube to observe their writhing responses, other mice recognized and responded to the other's pain. On witnessing their writhing comrade, the observing mice showed heightened sensitivity to painful stimuli. Though the scientists concluded that this is a "crude form of empathy," there was a further nuance. The mice typically only showed the contagious pain-response behavior if they knew the writhing mouse. Interestingly, males witnessing the sight of an unfamiliar writhing male mouse (a potential rival in nature) actually showed a drop in pain sensitivity. The rodents' differential responses show that they were not merely expressing an automatic fear reaction. This study not only demonstrates empathy in mice, but it also illustrates that they relate to each other as individuals, as with our own natural tendency to show greater concern toward friends and family.

One possible explanation for the more pronounced writhing response in the observing mice is that they were merely imitating the pained state of the other mouse, and not expressing any sort of empathy. However, when observing mice were exposed to a different source of pain—a radiant heat source—they moved away from it sooner in the presence of a writhing cage-mate than one that was not writhing.[31]

The irony of such a study is that our own empathy seems to fail us when we deliberately subject animals to a painful experiment. In addition to the painful injections, this study also involved the use of chemicals to permanently render mice deaf, or unable to smell. As happens in practically all laboratory studies of rodents, the mice, numbering several hundred, were also killed at the conclusion of the study.

One chimpanzee, Washoe (introduced in Chapter 6), showed that her kind is empathetic to other species. Recall that Beatrice and Allen Gardner of the University of Oklahoma taught American Sign Language to Washoe. When Beatrice became pregnant, Washoe became more attentive than usual and regularly asked questions (using sign language) about the baby. Washoe had had two pregnancies of her own, both of which had resulted in the infants' deaths. When Beatrice returned after an extended absence, Washoe acknowledged her return but was aloof. The teacher explained that she had had a miscarriage and signed to Washoe: "My baby died." Washoe looked at Beatrice and signed "Cry," then signed "Please person hug" as Beatrice was leaving.[32] This example shows how sophisticated emotions are not dependent on sophisticated language skills. In a similar incident, the bonobo Kanzi was grooming the hand of Bill Fields, a researcher with the Great Ape Trust in Des Moines, Iowa. When Kanzi got to where Fields is missing a finger, he pretended like it was there. After pausing, using his keyboard, Kanzi uttered, "Hurt?"[33]

In a study that used videos and photographs to examine emotionality in chimpanzees, subjects were required to select from a range of chimpanzee facial expressions to categorize emotional video scenes—such as favorite foods and objects, and (unpleasant) veterinarian procedures—according to their positive and negative valence. They spontaneously matched the videos to corresponding expressions according to their shared emotional meaning. For example, test subjects shown a video of chimps being given food selected a photo of a happy face, whereas the veterinary scenarios had them reaching for an unhappy face. This study indicates that chimpanzees process facial expressions emotionally, as do humans.[34]

Frans de Waal, mentioned in the previous chapter, was at a zoo once where keepers were trying to retrieve a monkey who had escaped into a tree. A group of chimps were watching from their enclosure. One of the males who was watching gave a little yelp and grasped the hand of another chimp as the monkey, shot by the zookeepers with a tranquilizer dart, fell out of the tree and into the net.[35]

Deference

We tend to show particular consideration and deference toward fellow citizens with physical or mental handicaps. So do monkeys. In *Our Inner Ape,* Frans de Waal describes a captive colony of rhesus monkeys he studied that included Azalea, a mentally retarded female born in their midst who had symptoms resembling human Down syndrome:

> Rhesus monkeys normally punish anyone who violates the rules of their strict society but Azalea was able to get away with the gravest of blunders, such as threatening the alpha male. It was as if everyone realized that nothing they did would ever change her ineptness.[36]

A free-ranging macaque group in the Japanese Alps included a female named Mozu who was born without hands or feet. Despite this severe handicap, Mozu was accepted and assisted by her comrades, and lived a long life, successfully rearing five offspring.[37] How do we reconcile such behavior with the conclusion of a research team in 1992 that rhesus macaques show no evidence of empathy?[38] One problem with their study was that it involved pairing macaques with a human partner. The setup of this experiment did not account for the possibility that the monkeys are less willing to show empathy across species lines, or the likelihood that the macaques harbored resentment toward humans for keeping them in laboratory cages. It is not easy to show that an animal possesses a capacity as nuanced as empathy, and it is a sign of progress that scientists are examining such questions. But excluding the possibility of empathy is much easier than showing that it exists; we should not be quick to conclude that an animal lacks such a complex emotion merely because a particular study design fails to show it.

We've seen that dogs react negatively to unfair treatment. Can they also show deference? While Finnish biology student Pauliina Laurila was studying animal behavior and ecology in England, she cared for homeless cats and dogs in her house. Her old Belgium shepherd male, Rommi, was in the house too, looking over things. Here's what she observed one day:

> I had rescued an orphaned kitten, about 2 weeks old. It was tiny, helpless and so alone in the world. Rommi was 11 years old at the time. I got the idea that the kitten needs the warmth of another animal, so while Rommi was lying on the floor I placed the kitten next to his belly. The dog didn't seem happy but let the kitten stay there. The kitten relocated to Rommi's front paw and fell asleep. So the kitten was sound asleep on the paw and guess what my dog did? Absolutely nothing. He hardly even breathed, he did not move a muscle! This happened on other occasions. Whenever the kitten fell asleep beside the dog he waited absolutely still.

It would be interesting to find out whether Rommi was motivated more by a desire to nurture the kitten or to please Pauliina. I asked Pauliina what she thought, and she felt that it seemed more likely Rommi was nurturing the kitten, for when she left the room without instructing Rommi to stay, Rommi remained where he was. Either way Rommi showed consideration for the needs of another being.

Hugo van Lawick noticed that the elephants he had observed on the Serengeti Plain behaved very gently not only toward each other but also with much smaller animals, such as baboons. When these creatures wandered among their feet, the elephants moved more slowly and cautiously, as if aware of the destructive potential of their own size and weight. Marie Galloway, an elephant keeper for twenty years at the National Zoo in Washington, D.C., describes an occasion when a helicopter flew too low. She found herself inside a protective huddle of elephants, a place usually reserved for calves.[39]

When a dog shows deference for a kitten, or when a rat forgoes food to relieve the pain of another, I think what we're seeing is the natural behavior of a socially adept animal. Caring about others is what keeps societies efficiently humming along. Pain is bad, no matter who is feeling it, and if it takes a bit of self-sacrifice to help another, that's the natural thing to do.

Democrats

Most human societies believe that decisions ought to be made democratically. We form committees, stage elections, and, in most cases, arrive at decisions either by negotiating until a compromise consensus is reached or by agreeing to a majority opinion. Until recently, decision-making in animal groups was thought to be autocratic—the leader decides, and all follow. Evidence has begun to accumulate that many animals also try to canvass a group's opinion to make democratic decisions. Recent studies by Tim Roper and Larissa Conradt from the University of Sussex find that when a group of animals makes a decision to move from one place to another—be it a flock of geese flying to another lake, or deer walking to another pasture—that decision is made only when a majority (typically about 60 percent) make intentions to move.[40] In the case of deer, an individual's "vote" is made by standing up. Among African buffalo, females make the decisions, and vote with the direction of their gaze.[41] Whooper swans use head movements. As we might expect of congregations of beings who are autonomous, not automatons, individuals in the group may have different interests and desires at a given moment, but they are willing to suppress those immediate wants for the greater benefit of staying with the group.

Democracy also occurs in social insects. From time to time, honeybees need to find new nest sites. How does a colony of twenty thousand bees make this decision? Although the hive has a matriarch in the queen, the plans for relocation are not autocratic but decidedly democratic. Bees fly out from the swarm to scout potentially good new nest sites, then return to recruit others who fly out to assess the site. The level of enthusiasm for prospect sites is conveyed in the intensity of the waggle dances performed by returning recruits. Every time a bee returns from a site, she makes about fifteen fewer runs than on her previous trip. So the number of recruits, and subsequent support for a site, can fade away if recruits disagree with a scout's assessment. A computer simulation of this "quorum sensing," starting with 150 runs and reducing by fifteen runs each time, ensures that the choices of error-prone bees are filtered out and that a favorable decision is efficiently reached.[42]

Good Deeds

In March 2008, a bottlenose dolphin named Moko at New Zealand's Mahia Beach rescued two pygmy sperm whales. The whales had become stranded in a narrow, shallow finger of seawater between a sandbar and the shore. Conservation Department officer Malcolm Smith had tried unsuccessfully to push the ten-foot mother whale and her five-foot calf back out to sea. When Moko arrived, Smith could hear the whales and dolphins making sounds, apparently to one another. Moko then led the whales about two hundred yards along a channel of water parallel to the shore, then she turned a right angle and escorted them out to sea.[43]

It seems likely that for every incident of this sort that is witnessed by humans, many more occur out of view. The article also reports that whales become stranded on Mahia Beach about thirty times a year, and that most incidents end with the whales being euthanized. This bit of information reveals much about our different attitudes toward nonhumans. Can you imagine reading that *Each year some thirty people get lost in Desert X and most end up having to be put down?*! Euthanasia is certainly better than being left to die a slow death, but surely human ingenuity and morals can do better than this. What about inventing an inflatable raft with a harness that can be slipped beneath a stranded whale, inflated, then used to pull the animal to safety? What is stopping us from intervening in a more systematic, determined fashion?

In the Tuli bushlands of Botswana where Gareth Patterson studied elephant populations, one of the elephants was maimed and her trunk was rendered useless by a poacher's snare. Gareth received reports from antipoaching guards that they had seen other elephants placing food into the mouth of their handicapped comrade, who was now surviving with just half a trunk. Gareth occasionally saw this elephant drinking, which she performed by reverting to the method of babies, who kneel down and suck water directly into their mouths.[44]

Cynthia Moss recounts the adoption of a one-week-old elephant calf by a family unit. The calf was orphaned when his mother was killed outside Uganda's Rwenzori National Park. Simon Trevor, a filmmaker, and Michael Woodford, a veterinarian, decided to introduce the calf to a family unit inside the park. When they released the calf near an elephant herd, the calf kept following the men. In desperation, they pushed the calf headfirst into a bush, causing the animal to trumpet.

This drew the attention of the nearby herd, whose replies sent the calf in their direction. The first elephant the calf approached was unfortunately a large bull, who knocked him flying with a swat of his trunk. The second was a nursing cow, who gently drew him toward her with her trunk and allowed him to nurse. The team tracked the herd the following day, and the calf appeared to be an accepted member of the family unit. There is some evidence that allomothering (nonmaternal infant care) in elephants is more prevalent during hard times—when it is most needed.[45]

Animals are not rigidly fixed in their evolutionary roles. As sociable, engaged, considerate beings they show flexibility in their behavior. Giving, cooperating, and nurturing are basic elements in life. As babies, we are the recipients of the largesse of another, and as we grow into adults we reciprocate. Life is sustained through the generations by the goodness of others.

Rich Experience

The main elements of animal sentience and coexistence are their sensory, perceptual, emotional, and cognitive capacities, which together engender a rich world of experience. The presence of others enriches the experience further, and gives rise to emergent properties: communication, sociality, and virtue. It is a mistake to think that morality originated with *Homo sapiens.* Being nice was adaptive long before we stalked the earth, and virtuous nature is beginning to garner the scientific attention it deserves. Let me also emphasize that the reason why I think it is important for us to be conscious of other animals' own forms of intelligence, awareness, and virtues is not to liken them to us, but rather so that we might realize that they have lives worth living.

Part III takes the information from Parts I and II and applies it to our relationship with animals. In the next chapter I want to address a question that Part II has been leading up to: Is life worth living for animals? This is a deeply philosophical question. It is also accessible to science. Too often we encounter depictions of wild nature as harsh and cruel, with animals locked in an earnest struggle for survival. In a discussion that includes examples of the benignity and sustainability with which animals conduct their affairs, I want to show that life for animals is very much worth living. In Chapter 10 I put human nature under the microscope, arguing that it is hypocritical to portray animal life as uncivilized in light of our ongoing history of violent conflict and our institutionalized indifference toward animals. The final chapter is about the path that humankind can, and really must, choose if we want our grandchildren to inherit a better world. For too long, an attitude that might makes right has governed humankind's relationship to animals. It is time our hearts caught up with our knowledge. Grounded in science and driven by ethics, I argue for an emergent, less selfish worldview that grants animals the respect and consideration they are due.

Part III

Emergence

There is a scene in the film *Monty Python and the Holy Grail* in which King Arthur encounters two serfs heaving clods of earth in a rural medieval landscape. The serfs are filthy and bedraggled, looking every bit the part of the uneducated underclass befitting their typecast. It becomes quickly apparent, however, that these serfs—Dennis and "Woman," are experts on social politics. King Arthur, simply wanting to know what knight lives in yonder castle, soon finds himself being lectured to on the injustices of a "self-perpetuating autocracy" and the "outdated imperialist dogma" that led to Arthur's position of absolute power.

When the woman asks Arthur how he became king (noting peevishly that she didn't vote for him), Arthur—accompanied by singing angels—describes having the sword Excalibur bestowed upon him by the Lady of the Lake rising from the bosom of the water. This sends Dennis into a diatribe against "farcical aquatic ceremonies" and "strange women lying in ponds distributing swords as a basis for a system of government." "I mean," says Dennis, "if I went 'round saying I was an emperor just because some moistened bint had lobbed a scimitar at me, they'd put me away!" Arthur, growing impatient, repeatedly tells Dennis to shut up.

This scene can be viewed as a scathing attack on monarchy, autocracy, dictatorships, and—to presage Monty Python's subsequent film, *Life of Brian*—religion. I see it also as a metaphor for the human–animal relationship. King Arthur is humanity. The serfs are the animals. We are their self-appointed lords, they are our servants. And, just as King Arthur rests his claim to superiority on a faulty foundation, so too is our dictatorship over the animals based on outmoded thinking and perpetuated by wobbly ethics.

Parts I and II of this book have been my effort to buoy the serfs, to speak up for the animals, to raise them (or our conceptions of them) from the muck. Our deepening understanding of animals in recent decades has outpaced our cultural, ethical, and legal views toward animals, which were formed largely in an earlier era. We can never revert to a Cartesian naïveté that would deny them a stake in their lives. Today it can be debated whether or not animals have souls (Descartes argued that only humans did). But there is no debate that animals are conscious beings, not automatons without feelings, as Descartes claimed.

But here's the rub. Our treatment of animals remains medieval, or arguably worse. Up to and during the Middle Ages, animals were mostly viewed as things that were created to serve human ends. Today, while that view is beginning to wane, we have factory farms, mechanized slaughterhouses, and laboratories that conduct toxicity tests on animals. As long as these institutions remain in place, our humanity lags behind our emerging knowledge of animal sentience. Humankind has begun to address, but it has scarcely begun to correct, the wrongs of our protracted imperialism toward animals.

How are we going to change? What forces are holding us back? By what means are humans going to come to our senses and adopt a new worldview of all sentient life?

CHAPTER NINE

rethinking cruel nature

"Early taxidermists exaggerated the ferocity of all animals that came to their table. In their hands, even a mole was poised to spring, its two teeth exposed."

J. M. Ledgard, *Giraffe* (2006)

Part of what sustains our imperialistic view of animals is an ages-old perception that *brute nature*—and by "nature" I'm referring to the lives of animals who live free in the wild—is unrelentingly harsh and cruel. My aim in this chapter is to challenge that perception. As we have already seen, nature involves a lot more than taking what one can get. Evolution favors those who live economically. Profligacy is discouraged, and cruelty is often inefficient.

When I was a boy, my family would sit down on Sunday evenings to watch the Walt Disney television special. It often featured live-action animal dramas, and my sister would sometimes leave partway through because she couldn't bear to see a coyote trying to avoid drowning in rapids, or lemmings making suicidal leaps from a cliff. These films were a strong influence on me. At age nine I was inspired by such a film to write an essay titled *The Adventures of a Salmon.* Here are a few extracts from it, verbatim, to illustrate a sensitive lad's formative take on the vicissitudes of life in the uncivilized realms:

> ...A mother salmon was swimming around in the water when a male salmon came swimming by. It was the matting season and of course the male swam towards the female and started to mate in the process

> of mating you would think that they are fighting but actually they are thoughrally injoing it. . . . There had to be many eggs because salmon have many enemies and all the eggs would be eaten up. After afew weaks or so there were about 15 eggs left. . . . By a month [the hatchlings] had fed a lot and grown big enough to recognize that they were salmon. . . . Suddenly a big object came swimming towards them all the little fish swam for their lives. But most of them got caught and eaten up it was a big pike. Only two fish managed to escape the fish. A few were injured and were lying on the bottom gasping for air. They would soon die and would just be a few bones on the ground. . . . [The survivors faced] a long and dangerous trip because finding and making a home was not easy.

That film made an impression on me. Like most children, I accepted without question that fish can feel things. But more to the point, my perception was that wild existence for a fish is hard and cruel.

Trials of Life

Virtually everyone knows the phrase "Nature red in tooth and claw," though few of us know its origin. It comes from a long poem by Alfred, Lord Tennyson published around 1850. The poem, titled *In Memoriam A.H.H.,* is dedicated to Tennyson's Cambridge University friend Arthur Henry Hallam, who had died suddenly from a brain hemorrhage. Tennyson was referring to nature's cruel indifference.

Tennyson's sentiments are understandable. Nature is blind to our personal hopes and feelings. She does not play favorites, and we may rightly feel anger or betrayal when a dear friend or family member is struck down by illness.

But there is a nefarious side to the broader meaning of Tennyson's alluring phrase. Whether he intended it or not, Tennyson's words reinforce an impoverished view of animal nature. It is the view that wild existence is mostly about suffering and hardship, and it is championed by some of today's most influential scientific thinkers. I refer to such luminaries as evolutionary biologists Richard Dawkins and George C. Williams, and philosopher of science Daniel Dennett. They tend to emphasize the misery and destruction to be found in nature, and neglect the positive aspects of wild existence.

Dennett alludes to animal nature as nasty and brutish, and endorses seventeenth-century philosopher Thomas Hobbes's similar perception of the state of nature.[1] Hobbes described life on earth as "solitary, poor, nasty, brutish, and short." Dennett invokes the easy target of "The Polyannian Paradigm, which cheerfully assumes that Nature is Nice," and recommends George Williams—a distinguished evolutionary biologist from the State University of New York—as "a powerful antidote."[2] As Dennett explains: "Williams points out that, in all the mammalian species that have so far been carefully studied, the rate at which their members engage in the killing of conspecifics [members of the same species] is several thousand times greater than the highest homicide rate measured in any American city."[3]

Oxford University's Richard Dawkins—perhaps the world's best-known popularizer of science—is no less sparing of nature than Dennett or Williams. In his 1993 book *River Out of Eden: A Darwinian View of Life,* Dawkins writes: "The total amount of suffering per year in the natural world is beyond all decent contemplation. During the minute it takes me to compose this sentence, thousands of animals are being eaten alive; others are running for their lives, whimpering with fear; others are being slowly devoured from within by rasping parasites; thousands of all kinds are dying from starvation, thirst and disease."[4]

I admire the contributions these men have made to our understanding of evolutionary biology and I share many of their fundamental viewpoints. I completely agree with George Williams, who explains in an essay titled "Mother Nature Is a Wicked Old Witch," that planet Earth was not formed to be an agreeable place for humankind (or any other creature) to live.[5] We and all other organisms have had to work hard (in the evolutionary sense) to adapt to the Earth's geological belchings and thrustings, harsh temperatures and climatic regimes. In Williams's words, "organisms adapt to their environments, never vice versa."[6] But, as I hope to have shown in Part II of this book, the interplay of animals with brains and feelings is very different from the functioning of an inanimate planet without a brain or feelings. As we've seen, the evolution of societies has spawned a great deal of cooperative, virtuous behavior.

What drives learned scientists and philosophers to belabor the notion of "cruel nature"? There are, I believe, two deep cultural rationalizations at work. First, the idea raises humans to a higher plane and puts us on the moral high ground. Compared with savage nature (so the thinking goes), humans are civilized and moral. An identical rationale was used to defend our subjugation and mistreatment of other human beings labeled as savages—American Indians, enslaved Africans, and others. Today, for the most part, we have removed the arcane designation of "savage" and now treat them as fellow humans. Second, if nature is cruel and harsh, then we can claim our own savagery toward animals as merely part of the natural process. Cruel nature absolves us of any guilt for treating its denizens cruelly. Thus, cruel nature provides a sweeping palliative for our own moral shortcomings. Whatever we do, we are justified.

The view of wild nature as an earnest, relentless struggle to eke out one's existence in an indifferent world is a cynical misconception. Evolution is as indifferent as a stone, but animals are not. Evolution is a process, and animals are a product of that process. We animals have emergent properties: conscious awareness, intelligence, emotions, societies, culture—all those things discussed in Parts I and II. Sure, evolution shows trends. Animals, for instance, have tended to become larger and more complex since the primordial soup came off the burner, so to speak. But, being blind, evolution has no goal in sight.

Animals, by contrast, do have definite goals, such as getting food, finding mates, and seeking comfort. As we saw in Chapters 7 and 8, wild nature isn't a selfish free-for-all. Living together has advantages, and it requires getting along by being cooperative and considerate. Of course, there is strife and conflict in nature, but nature also is respectful, benign, and often beneficent. As I argued in my book *Pleasurable Kingdom,* animals don't merely exist; they experience their lives richly. For them, as for us, life is worth living.[7]

Evolution often favors the triumph of benefits over deficits, of affirmation over negation. Animals evolve not just to avoid dangerous things, but to seek and gain useful and desirable things. Their daily activities are a source of reward and satisfaction. Because seeking food, securing shelter, and establishing social partnerships are essential to survival, it doesn't logically follow that there is no joy in these activities. On the contrary, their very importance is why the activities of eating, getting comfortable, and interacting with others are routinely pleasurable. Pleasure-seeking is adaptive. In a carrot-and-stick world, pleasure is nature's carrot, the treat we aim for.

The same holds for genetic survival—a concept to which most animals are oblivious. Given its indispensability to Darwinian fitness, sex has evolved to be rewarding for animals. In his 1999 book *Biological Exuberance,* biologist Bruce Bagemihl detailed the breadth and scale of currently known nonprocreative sexual behavior in mammals and birds. Many hundreds of species have been found to engage in one or more erotic activities that in strictly genetic terms are wasteful and inefficient, such as same-sex mating, masturbation, manual or oral stimulation of others, and mating outside a prescribed breeding season.[8] Animals do not just go through the motions—they are highly motivated.

How we choose to view wild nature strongly influences our relationship to it. If we view animals as savages living in a relentlessly harsh world, we unconsciously place ourselves on a higher plane. But when we see that animals have lives of value, we may recognize that we have a responsibility to show respect and to allow them space to pursue their existence. However we view nature, as creatures of conscience, we can choose to cause more or less suffering to animals. Even the grimmest view of nature can be seen to warrant kinder treatment of animals, for we may spare them additional suffering.

Peddling Predation

Like it or not, we are drawn to violence and drama. Switch on the television, open the daily newspaper, or peruse the weekly newsmagazine and you are likely to find a preponderance of violent conflict, turmoil, and disaster. A million successful bus or car rides are not news. It is the one that ends tragically that we are likely to hear about. *If it bleeds, it leads,* goes one of the aphorisms of journalism. Our fascination with carnage and bloodshed extends to our popular conceptions of animal nature. Ask someone to name a species of shark; if they know any at all, it's most likely to be one of the few species that have made rare attacks on humans, such as great white or tiger sharks. Most of us have never heard of the benign species which comprise the great majority of the 360 living shark species, such as the cat shark, the crocodile shark, or the spiny dogfish. It's the same with bats. Few of us, I suspect, can name anything other than "the vampire bat." There are actually three species of vampire bats, which in themselves are far more benign than their reputation (recall the discussion of caregiving in common vampire bats from Chapter 7). Only a biologist is likely to be able to name any of the remaining thousand-odd species. Piotr Naskrecki, director of the Invertebrate Diversity Initiative of Conservation International and author of *The Smaller Majority* (2005), bemoans the fact that negative myths about benign animals abound in almost every culture. For instance, there are only two venomous lizard species worldwide—the quite nonaggressive Gila monster and the beaded lizard of North America and Latin America. Yet virtually every culture considers at least one local species of skink, gecko, or chameleon to be extremely dangerous or venomous.[9]

Nature documentaries and "Animal Face-Off" videos on Animal Planet do their part to reinforce a view of wild nature as a constant struggle. In a documentary film on the lives of a herd of wildebeest, a disproportionate amount of footage is taken up with a twice-yearly perilous river crossing during migration, when many (though a small fraction of the whole) fall victim to drowning, submerged crocodiles, or ambushing leopards. Polar bears may spend the bulk of their winter waking hours hunting for seals on the pack ice, but they normally go days between kills, each of which lasts only a few seconds. Yet, the kill is the focus of the cameras. Reindeer are faced with *a race against time* on their long annual migrations to summer and winter feeding grounds; time-lapse photography accentuates their harried progress and shows them streaming across the landscape, making brief stops to refuel.

When I watch these programs, the message I am left with is that every living, breathing creature is faced with an earnest struggle and is in imminent danger of death by starvation or thirst, or at the claws and jaws of a lurking enemy. It is the "animal kingdom" version of television news producers' obsession with extreme weather and disasters.

Emmy Award–winning wildlife film producer Kathryn Pasternak told me of her struggles to keep a documentary film on hyenas from being edited to accentuate violence. She managed to retain her preferred title of *Hyena Queen* for the international version, but the version that aired on U.S. television was renamed *Hyenas at War.* (Ironically, *Hyena Queen* sold well, while *Hyenas at War* did poorly, perhaps owing to the American public's fatigue with war.)

Just how predaceous is nature? The flow of energy through an ecosystem dictates that predators are much rarer than potential prey. Calculations of the energy dynamics of living systems find that even the most efficient ecosystem loses about 98 percent of its energy at each level of a food chain. The main causes of this energy drainage are heat production, the energy required to run an organism's body, and energy required for movement. That means, as American ecologist Paul Colinvaux explained, "all the insects in a woodlot weigh many times as much as all the birds; and all the songbirds, squirrels, and mice combined weigh vastly more than all the foxes, hawks, and owls combined."[10]

Predators and prey have conflicting interests, but the rules of energetics (the study of the flow and storage of energy) preclude gluttony. There are no obese lions on the Serengeti, although the irritating presence of scavengers who waste little time in taking the leftovers ensures that there are often very full lions. There are few if any fat predators anywhere in the wild. There are good reasons for this. Catching prey is hard, often dangerous work. Killing a glut of prey is also a poor strategy because the meat will spoil (except in cold regions where some large predators will stash and return to a kill), and it is vulnerable to scavengers with a keen sense of smell.

Beyond the energetic limitations inherent in food chains, being a predator isn't an easy profession. Obviously, would-be prey do not make themselves easy to catch. They may be evolved to escape detection by means of camouflage, or to escape capture through speed. They may situate themselves out of reach, such as high in the treetops against large terrestrial predators, or in narrow burrows. They also may signal their undesirability to predators. Many species of African ungulates (hoofed animals) will bounce repeatedly high in the air, lifting all feet off the ground simultaneously—a behavior termed stotting or pronking—in the presence of a predator. The behavior is believed to notify predators that they have been detected; it may also act as an advertisement of fitness, as if to say *don't bother trying to catch me because I'm fit and fast.*

A study of predation incidence at eight colonies of prairie dogs found that while predators hunted there quite frequently, only about 3 percent of hunts were successful.[11] Predation rate on savannah antelope and certain ground-based monkeys is reportedly around 6 to 10 percent, meaning that an individual's chances of avoiding predation in a given year is between about 90 and 96 percent.[12]

Not Doomed by Handicap

Another way that our view of life for wild animals is made harsh is in the belief that an illness or an injury is a death knell. As such, nature has no room for anyone running on less than all cylinders. The injured and the sick are quickly weeded out, so we're taught. You will mostly likely have watched a nature documentary in which the narrator laments the doomed fate of an injured animal. Certainly, a lion with a broken jaw or a bird with a broken wing probably doesn't have long to survive in the wild. But it doesn't follow that the slightest disability spells death from starvation or in the jaws of a predator or from failure to compete with one's fitter comrades. In the real world, untainted by human bias, many animals survive and thrive with handicaps.

While visiting historic Fountains Abbey in Yorkshire, England, in January 2005, I found myself among a large flock of jackdaws foraging among grass kept well trimmed by a mixed herd of about five hundred deer. There were about a hundred birds, loosely scattered, each strolling about with their heads down, stopping intermittently to pluck up a beetle or investigate the underside of a leaf. I noticed one individual nearby who had no right eye. There was no sign of recent injury, just a small dark depression where a jackdaw's normally prominent grayish-white eye would be. I guessed this bird must either have been blinded in a mishap or been born that way—probably the former because of the difficulties of learning to fly and forage as a fledgling. Yet she—let's assume she is female—looked in splendid condition. Her body was plump and her feathers shiny. Even her head movements gave no clue to the handicap, which conventional science wisdom would hold to be serious for an animal that relies heavily on bilateral vision to fly and to detect dangers. But this jackdaw lived gregariously, benefiting from the eyes of her comrades to detect potential danger on her blind side. There was no physical or behavioral sign that the bird was either sick or infirm. We might say that she defied the hackneyed claim that nature soon weeds out the weak and infirm—except that despite her imperfection she was neither weak nor infirm.

A month after I watched the one-eyed jackdaw, an old friend emailed me from Ontario, Canada, to say that she had watched a one-legged peregrine falcon feeding from a roadside carcass. She also noticed that the healthy left leg was banded. She called the local naturalist club, and learned that the bird was known to them. Dundas, as they called him, had hatched from a nearby nest site the previous year. He had begun life with both feet, and he was the smallest bird of the brood. Later, Dundas was observed making a kill. As of December 2006, Dundas was being observed regularly in Kitchener, Ontario, where he had attracted a mate.

Len Howard witnessed many instances of nestlings and fledglings who recovered from injuries that appeared to be hopeless. Parent birds continued to feed fledglings weakened by injury, at times giving them special attention by continuing to feed them after nest-mates had become independent.[13]

For animals who live in societies, like baboons, the prospects of survival with handicaps are probably better because of the buffering effect of living among others. Job, an adult male baboon in a population studied by Robert Sapolsky, was handicapped by palsies and seizures, undescended testes, and some kind of fungal infection. He remained near the bottom of the troop hierarchy, but he was a survivor despite his disabilities. Another baboon in this troop, Miriam, suffered a broken arm as an infant when the adult male Nebuchanezzar grabbed her one day to discourage a more dominant male's attack. She limped into adulthood, but she didn't die. Neb himself went through life missing an eye. Baboons will befriend and help others in their group, and they benefit from the collective vigilance of others. As we saw in Chapter 8, monkeys and apes who are physically or mentally handicapped may lead normal, even productive lives when they are accepted into their social networks.[14]

Limpy, a male wolf wounded in his first year of life in a fierce fight with a neighboring pack, not only survived, but lived another eight years. In 2002, he traveled from Yellowstone National Park, Wyoming, and was caught in a leghold trap near Salt Lake City—the first wolf sighting in Utah in seventy years. Trucked back to Yellowstone by the U.S. Fish and Wildlife Service, Limpy quickly reintegrated with his home pack and his injured foot recovered. He was still hale and hearty six years later when he was shot dead, along with two of his companions on March 28, 2008, the first day wolves lost their protected status under the U.S. Endangered Species Act.[15]

These accounts illustrate the tenacity of wild animals and their capacity to survive—and their communities' willingness to help them live—despite debilitating injury. They also belie popular perceptions that the slightest disadvantage spells doom for a wild creature. For many species the survival bottleneck is narrowest during the early stages of life, when naïve young animals have much to learn and are more vulnerable to predation, mistakes, and injuries. In the springtime I often walk past recently fledged robins and other birds who seem rather bewildered with the new surroundings outside the nest and make little effort to widen their distance from me. A few days on, those who survive are savvier. With the transition to adulthood, survival skills are honed and the prospects for longevity look good.

Violence and Dominance

Nature doesn't countenance violence without good cause—it prefers more peaceful paths. Arabian oryxes kneel down before engaging in tussles with their long, backward-curving horns. Many snake species wrestle among themselves, but their fangs are not deployed. Mule deer bucks refuse to strike "foul blows" or "cheap shots" when their opponents expose the unprotected sides of their bodies. Nature evolves many means to avoid conflict and subvert death. Territorial calls, scent markings, and other displays establish spatial boundaries that are usually known and respected by locals. Social groupings are composed of individuals who know one another and live mostly peaceably.

Our changing understanding of baboons illustrates how traditional negative views of animals can give way to more optimistic ones. Robert Ardrey described the baboon as "a born bully, a born criminal, a born candidate for the hangman's noose."[16] Ardrey and other popularizers of animal behavior during the 1960s and 1970s based their interpretations on early studies, such as those of Sherwood Washburn and Irven

DeVore, who worked for ten months in 1959 and 1960. The picture that emerged from this study was that the structure of baboon society was maintained by an elite clique of brutal, domineering males.[17]

Subsequent studies, of longer duration, have revealed a quite different picture. As Cynthia Moss summarizes: "A baboon troop is not held together by fear and attraction to despots, but by a complex matrix of family relationships and social bonds. The baboon social system described by the new research may not support the popular author's theories of male status and dominance, but it seems to be a logical, successful, and far less rigid way of life."[18] Shirley Strum, another veteran of studies of free-living baboons, thinks that the idea of a "dominance" system—in which hierarchies are rigid and enforced through intimidation by large males—is too simplistic for a society as complex as that of baboons.[19] Robert Sapolsky believes that while the alpha male forms the troop's epicenter, it is the females who are its bedrock.[20]

Elsewhere, dominance is proving to not always be the best strategy. Sometimes there can be more to gain from maintaining a subordinate

Baboons groom in a suburb of Cape Town, South Africa. (Photo by the author.)

position than from risking violent conflict, which risks injury or worse. Rats have a promiscuous mating system; male rats engaged in fighting over access to a female may open the way for another male to mate with the female.

A recent study of a wild colony of jackdaws found that dominant males had lower reproductive fitness. These males consistently produced fewer fledglings, whose survivorship at one year old was lower than the young of subordinate birds. Females who mated with dominant males were also in poorer condition and laid smaller eggs.[21] The authors speculate that the colony's high density led to high testosterone levels in males, who become less nurturing of their mates and offspring.

A study of Atlantic blue-eye fishes also found that dominance had drawbacks. Females did not always prefer dominant males over subordinate ones, nor was mating with males based on dominance a guarantee of maximal benefits to females.[22] In another fish, the sand goby, females have been shown to prefer good fathers, who are better at taking care of the eggs, over dominant males, and to have better reproductive success as a consequence.[23]

Nature's Grimmer Side

My aim is not to represent nature as roundly fluffy and pleasant. Nature certainly contains its share of pain and hardship. Many creatures must kill others to eat. Others are parasites and live off their hosts. In some circumstances, animals may cannibalize one another, kill their own or another's babies (infanticide), or their siblings (siblicide).

It is standard operating procedure for male lions to kill any young cubs once they take over a new pride. In this way, they will not waste energy raising cubs with some other male's genes. Interloping male lemurs also sometimes kill young, which brings females into mating condition. Among a long list of species known to commit infanticide are wild horses, dolphins, langurs, coots, wattled jacanas, rats, mice, hyenas, chimpanzees, gorillas, and humans. Dorothy Cheney and Robert Seyfarth report that at least 53 percent of the infants born in their study of wild baboons in Botswana were the victims of confirmed or suspected infanticide.[24]

Infanticide and cannibalism can result from unnatural conditions imposed by captivity. The confinement and stress of factory farming systems, in which individuals are crowded into small spaces and are unable to engage in natural behaviors, are often implicated. Cannibalism among adult layer hens, for instance, can be linked to social disruption, bright lighting, onset of sexual maturity, overcrowding or close confinement, large group size, and the presence of sick or dead birds.[25] Close observational studies of a population of reintroduced Przewalski's horses found that infanticide was only committed by stallions who had grown up under unnatural social conditions (i.e., in captivity). Furthermore, killings occurred only during the first years following reintroduction, when social relationships were unstable and the horses nervous. No stallion of the second generation killed a foal.[26]

Infanticide is inevitably costly to a mother who invested time and energy to gestate and nurse an infant. One way to cut costs is to abort or reabsorb a fetus when the prospects of its survival are poor. Some pregnant rodents will do this in the presence of a new male, "probably because he would likely kill her litter anyway." This ability is named the Bruce Effect after the biologist Hilda Margaret Bruce, who first described it in 1959.[27] If a male mouse can commit infanticide, it doesn't follow that he cannot be a good parent. Like mother mice, fathers clean their pups, retrieve them if they climb out of the nest, defend them against attack, and escort them during early exploratory forays.[28]

Sibling aggression occurs in a minority of bird species, and in some, such as boobies, kittiwakes, guillemots, herons, egrets, and osprey, it can result in siblicide. In so-called obligately siblicidal species such as the masked booby, the last-born chick nearly always dies. Siblicide is theorized to be an insurance policy. In times of plenty, both chicks get enough food and the parents may raise two chicks instead of one. In leaner times, a second chick becomes the back-up should the first chick falls ill or dies. Barring these circumstances, the stronger chick may effectively starve the smaller one who hatched later by monopolizing the food supply. Or the younger may be the victim of physical aggression by the other chick, including being pushed out of the nest. It's a heartless system, but for good reason it is relatively rare. Fighting among nest-mates risks injury to the combatants and consumes energy that might be more profitably relegated to growth and maturation. When there are more chicks in a nest, there are more defenders to repel small predators, parents are less likely to abandon the nest, and the insulation of other warm bodies reduces risk of hypothermia by about half.[29]

Far more common and widespread than these lethal interactions are nonlethal ones. Parasitism is one of several life strategies involving the intimate association of one species with another. Because parasites benefit at the expense of their hosts, the parasite/host relationship can be described as plus/minus (+/-). A relationship in which both species benefit is termed a mutualism (+/+; the term was introduced in Chapters 6 and 7) and one in which one benefits and the other is apparently neither harmed nor helped is referred to as commensal (+/0). Readers should take note that nature is bereft of -/- relationships. Evolution—nature itself—would not countenance a negative/negative relationship. Spite may occur rarely in highly social species. Otherwise someone virtually always wins in life, even if there is often a loser.

Parasitism is a successful survival strategy. It is likely that there are more parasitic species on earth than pursuers of any other strategy.[30] This should not be too surprising, given that a single host can harbor many different parasite species and that parasites are in turn parasitized. In fact, all viruses are obligate parasites, and viruses outnumber all other organisms. Most of these viruses infect bacteria, many of which, as we saw in Chapter 8, happily reside within our bodies.

It is rarely beneficial for parasites to kill their host; a dead host isn't going to be a source of food or shelter for much longer. In most cases there is an adaptive balance between parasite and host, even though the host (if given a choice) would prefer not to be the host. By definition, a "parasitic" organism that eventually kills its host is not a parasite, technically speaking, but a *parasitoid.* Most parasitoids are tiny wasps that target other insects.[31]

The success of the parasitic life strategy need not paint nature as harsh and cruel. Many parasites go unnoticed by their hosts, who continue on with their lives with no awareness of the parasites they harbor. We are all carriers of an infinite number of parasites—some as minuscule as the mites living on your eyelashes. And as we'll see, parasitism can spawn benefits for the host as well.

The interactions between a tiny mite and its moth host illustrates the restraint built into these types of relationships. *Dicrocheles* mites live in the ears of moths. The ear of a moth is a dark, warm, and moist and makes an excellent niche to sip insect blood and raise a family. Unfortunately for the moth, the mites make their pad comfy and remove a few bits of old "furniture," deafening the moth in the process.

The moths are able to detect the echolocation calls of hunting bats and need to be able to escape their predators, which they cannot do if they're deafened. A grim view of nature might emphasize the selfish and cruel aspect of the mites' infestation. But a deaf moth who falls prey to a bat also spells the demise of its mite squatters. Presumably, this is why mites actively avoid migrating to the second ear. Even though the mite colony can get crowded at its peak, and the other ear is but a few steps away, the mites somehow know that ear number two is off-limits.[32]

Parasitic relationships can also lead to unforeseen benefits. Tiny wasps destroy fig flowers when they lay their eggs in the developing fruit. But these same wasps are important pollinators of the figs. The fig tree pays a price for pollination in terms of lost seeds. Enter player three. She is a different wasp—let's call her "B." A parasitoid of the fig wasp, she inserts her ovipositor through the flower to where the fig wasp larvae reside; when they hatch, her larvae eat the fig wasp eggs or larvae. But her ovipositor is only long enough to reach fig wasp larvae in shorter flowers or on the outer reaches of long ones. This may seem like a problem to the casual observer, but the fig trees have adapted by providing flowers of variable length. This three-way accommodation stabilizes the impacts of the fig wasp—sustaining the pollinator mutualism while helping to keep the fig wasp's parasitic effects in check. Incidentally, scientists had previously thought that parasite B was detrimental to both pollinator and tree.[33] Fig tree/wasp associations are among the most thoroughly studied, and it is likely that there are many other intricate plant–animal associations in nature that blur the line between a parasite and a symbiont. Some biologists theorize that many present-day mutualisms between organisms began as negative, or pathogenic, relationships.[34]

The interplay between parasitism and mutualism can get more complicated still. Oropendolas, charismatic, crow-size birds of tropical

Central and South America, often build their nests next to those of hornets or bees. These stinging and biting insects discourage monkeys, who sometimes take eggs from oropendola nests. Oropendolas also tolerate nest parasitism by giant cowbirds. Why? Because cowbird chicks pay their way by fending off parasitic botflies. Cowbird chicks will pick off botfly eggs and larvae from the oropendola chicks. Oropendolas appear to recognize cowbird eggs as not their own, for they will remove them if their nests are built near a hornet or bee nest. The stinging and biting insects also repel botflies, leaving no need for a cowbird chick's fly-removal services.[35]

Nature includes violence and suffering, but life isn't all tooth and claw. The authors of a book titled *Evolution: The Four Billion Year War* (1996) chose a violent metaphor for encapsulating life on earth, and they give prominence to Darwin's phrase "the struggle for existence." They go on to say, however, that cooperation is often a better strategy than conflict. As we saw in Chapter 7, animals who behave in concert with others often have better outcomes than if each acted selfishly. And while nature's laws are blind to suffering and morality, complex life has emergent properties that can mitigate nature's harsher tendencies. The cooperative benefits of social living trump selfishness, and virtue tends to be a better strategy than vice or violence. The intricate, often highly evolved relationships between hosts and their parasites also belie a simplistic interpretation that they are only harmful. To be sure, hosts would rather go without parasites. But the success of parasites has driven emergent host adaptations and has also led to secondary mutualisms with third parties.

Life, Worth Living

In a book titled *What Evolution Is* (2001), the great evolutionary biologist Ernst Mayr pointed out an inexorable fact of life: on a finite planet, assuming equilibrium in an ecosystem, for every set of parents there will be, on average, two young produced who survive to reproduce.[36] Since we think of animal life as being hardwired to spread our DNA far and wide, at first glance, this seems like a lot of work for little genetic benefit. All those salmon eggs and only two ultimate survivors. A lifetime of clutches from a mother robin, and only two to carry on the tradition. An adult female house mouse can produce up to ten litters of between three and fourteen pups in a year (if she herself survives that long). Yet, on average, only two will bear pups of their own.

Let me put these facts into perspective. The parent-offspring-replacement relationship contains assumptions that will err on the grim side of nature. Mayr knew this, which is why he included that clause *assuming equilibrium in an ecosystem.* For one thing, not every adult reproduces during its lifetime. In many species, younger individuals forgo breeding themselves, instead remaining with the parents and helping to raise their offspring (see the discussion of cooperative breeding in Chapter 7). In others, many individuals have no chance to mate. In harem-mating species, for example, one male dominates a large number of females and the remaining males must look on admiringly, and seek other ways to relieve sexual tension. Another assumption is that populations do not expand. In years of plenty they often do expand.

I think we should also question the common, tacit assumption that Mayr's one-to-one ratio somehow makes life not worth living for all those who die without reproducing. This is a matter mostly overlooked in discussions of nature. Reproducing may be the only point of life from a genetic perspective, but for conscious animals that feel things, there is much worth living for. Life for many animal species brings pleasures, even if it is foreshortened, or not graced by offspring.

In all vertebrates, and probably all animals, most of the dying happens early in life. For mammals, mortality is highest in the first year of life. After that, survival odds increase considerably. For African antelopes known as dikdiks about 50 percent of young reach the age at which they leave their natal territories, and many more of these are killed by predators during this relatively risky period of dispersal. Adult dikdiks, in contrast, have about 80 to 90 percent yearly survivorship. A study of giraffes in Nairobi National Park estimated that while about half of all calves die in their first three months of life, beyond that, overall yearly survivorship was about 87 percent, leading to an average life expectancy of about ten years. Cynthia Moss reported that in Manyara, Kenya, where conditions are favorable, annual elephant mortality was 10 percent for first-year calves and thereafter 3 to 4 percent, the same as for adult elephants.[37]

Predators seem to have a harder time surviving to adulthood, and they are especially susceptible to early mortality. Securing food tends to be more challenging both in terms of parenting burdens and of the young having to learn this indispensable part of survival. In George Schaller's landmark study of lions, 67 percent of the seventy-nine cubs born of the Seronera and Masai prides died: thirteen were killed by predators, fifteen died of starvation, and twenty-five from unknown causes. Schaller reckoned that between one-third and one-half of cheetah cubs die between the ages of about five to six weeks and three to four months, after which time mortality is low.[38] About one-half of leopard cubs don't reach adulthood.

Once animals get over the perilous hump of childhood, life expectancy rises. Elephants may live up to about seventy years. John Goddard, who studied black rhinos in Kenya in the 1960s, estimated their maximum longevity at forty years. Hans Kruuk estimated the oldest hyena from skulls he examined in the wild to be sixteen years. Wild female polar bears can live as long as thirty-two years; the oldest recorded male was twenty-eight. By analyzing chemical traces in the eye, and the presence of century-old harpoon tips embedded in the blubber of four bowhead whales killed by Inupiat Eskimos in northern Alaska, researchers from the Scripps Institution of Oceanography in La Jolla, California, in 2005 were able to estimate the whales' ages. The "youngest" was thought to be about 135 years old, and the oldest somewhere in the vicinity of 211 years. These whales were all in good health at the time they were killed. An analysis of seventy-five narwhals from Greenland using the eye chemical technique estimated the maximum age of a female at 115 years.[39]

It is commonly assumed that life expectancy in the wild is much shorter than in the relative safety of captivity. This is misleading, as it gives a false sense of both the perils of living free and the relative safety of captivity. According to Rob Laidlaw of ZooCheck Canada, captive life expectancy records are usually based not on average but maximum age records from zoos.[40] A recent article published in the journal *Science,* for instance, showed the median life expectancy of zoo-born African elephants in captivity was seventeen years, compared with fifty-six years in an animal reserve.[41] Elephants in zoos suffer foot problems, herpes, tuberculosis, and infanticide.

Finally, wild creatures are commonly thought to live on a razor's edge of survival. Such an existence would leave little time to waste on leisure activities. We commonly assume that life's projects are performed in earnest by wild animals, as if they never have any opportunity to do anything pleasurable. Many animal populations do appear to contain the maximum number of animals their habitat can sustain—this is called carrying capacity. That means there is, on average, just enough for everyone; it does not mean that getting one's share requires foraging around the clock. Furthermore, the pursuit of survival behaviors is not drudgery for animals. No view of wild nature is more impoverished than the belief that animals are unable to take pleasure in their daily activities.

Scientists studying wild black-tailed deer in Alaska concluded that the animals were living at the limits of their environment's carrying capacity. The implication would be that the food supply was stretched to the limit. Yet the animals ate more than they needed in all months except February. Predictably, the summer's glut of food was critical to winter survival, when the animals were unable to meet their energy needs. Even in cases like this, it is wrong to conclude that life has no dimension beyond a constant struggle to survive.[42]

The existence of extravagance is a measure of the degree to which animals are not leading lives of subsistence on the borders of life and death. Luxury is the product of excess. There are many extravagances in nature, especially among birds. The peacock is merely the best known of many birds that grow long tails and sprout showy plumages that limit their mobility and make them more conspicuous to predators—all in the interests of attracting a mate. In many wren species, the male builds several covered platform nests, from which his mate chooses just one to lay her eggs in, although the others may be used as a bad-weather shelter. So profligate are the bowers built by male bowerbirds to impress females that for decades researchers thought they were the work of diminutive forest-dwelling people. They couldn't bring themselves to believe the birds had done them.[43]

Studying animal behavior for nearly thirty years has given me a view of wild nature unlike the one we're taught to believe. It is time we stopped viewing wild nature as a constant, joyless struggle. We cannot know with precision just what it may feel like for a lion to chew at a fresh zebra carcass, or for a giraffe to stretch to her maximum height and coil her long tongue around a sprig. We cannot say for certain that a mouse relishes the taste of a beetle, or that a duck enjoys an afternoon nap. I don't know what a cowbird is thinking or sensing when he perches in a cluster of his own kind for thirty minutes at the top of a tree outside my window. What I do know with a fair degree of certainty is that their lives are not confined to a struggle for survival. However our existences compare with those of other animals, I cannot imagine that theirs could have no intrinsic worth. In the words of Albert Camus: "No one who lives in the sunlight makes a failure of his life."[44]

CHAPTER TEN

homo fallible

The wild, cruel beast is not behind the bars of the cage. He is in front of it.

—Axel Munthe, Swedish physician and author (1857–1949)

How often have you heard a violent criminal described as having behaved "like an animal?" "We may be hunted like animals, but we will not become animals," intones the narrator of a promotional trailer for the 2007 film *Defiance.* Recently, I awoke to a National Public Radio report on the discrimination of the dalits, familiarly known as the "untouchables," the lowest rung on the social ladder in south Asia, most notably India. In an interview, one of the Indian campaigners who champion the dalits' cause commented on the widespread cultural prejudices they face: "We don't consider them human beings, we consider them less than pigs and dogs."

Statements like these reinforce an age-old, deep-seated prejudice: that humans are better and more worthy than nonhumans. By subjugating a class of humans to the level of animals, we not only debase the human, we debase the animal.

How valid is this view that humans are superior? We may naturally favor humans over nonhumans because we are members of the human race. But what might we conclude from an unbiased assessment? In the previous few chapters, I have attempted to describe animals and portray wild nature in a more forgiving light than we usually place them. The current chapter turns the scrutiny toward our own species. How well does humankind stack up next to the "uncivilized hordes" we collectively refer to as animals?

The Unnatural Scale

One of the most pervasive ideas in human thought is that *Homo sapiens* is the pinnacle of life—that we are the pot of gold at the end of evolution's rainbow. Christianity teaches that we are created in God's image. The Reverend Dr. Jerry Bergman of the Institute of Creation Research writes: "For beauty and sheer simplicity of line, it [the form and function of the human body] is unmatched. As a machine, the human body is the pinnacle of God's work."[1] Humans "are the pinnacle of creation... because we are able to receive direct revelation from God and [we have] free will with which to put it into practice," writes Muslim scholar C. T. R. Hewer in *Understanding Islam*.[2]

These sentiments are comforting and reassuring for our need to feel special. And we are special, no doubt. But in the revealing light of scientific knowledge and evolutionary understanding, the teleological conceit that humans are what evolution has been working toward is simply and utterly false.

Stephen Jay Gould, one of the twentieth century's most celebrated explainers of evolution, sought to discredit the common presumption, especially prevalent among creationists, that evolution has a purpose, and that life forms proceed inexorably from less to more complex forms. He termed this pervasive misconception the "progressivist bias."[3] In its most pious, anthropocentric form, the progressivist bias views evolutionary history as advancing toward human beings as the highest life form.

The progressivist bias has ancient roots, going back some 2,300 years. The *scala naturae,* the "scale or ladder of nature" or "great chain of being," is a philosophical view of nature attributed to Aristotle and dating to the fourth century BCE. *Scala naturae* places nature in a hierarchical context. Inorganic objects such as rocks and bodies of water occupy the lowest levels, followed by plants, then "lower" animals (invertebrates) to "higher" animals (vertebrates), then to humans at the pinnacle of all life forms.

Though the term *scala naturae* is not in popular use today, it has been and continues to be one of the most influential ideas affecting humans' relationship to nonhumans. *Scala naturae* misconstrues humans as the ultimate manifestation of life.[4] A related thread is the teleological thinking that mankind is the goal of evolution: All of evolution has been leading up to man, and all other life forms are merely stepping stones toward this end. This anthropocentric idea is enshrined in the widespread religious doctrine that man is created in the image of God. Creationist and intelligent design views grant human supremacy unquestionable status by skipping evolution altogether—we were created and the other beasts were put there for us.

One of its implications is that there is an end point to the process. We evolve toward a perfect ideal, and once a species has been wrought to perfection, the changes stop. Another conclusion is that evolution leads inexorably to more complex life forms. Thus, a creature with more complexity of form and function—a spider, say, as compared to a sponge—is presumed to be more highly evolved.

Science shows that there are many problems with this premise. One is that evolution—even for the simplest and most "primitive" of organisms—is ongoing. A sponge is no less subject to evolution's whims than a spider or a dog. If sponges have changed less through time's passage than spiders or spaniels, it is because their design is less vulnerable to environmental upheavals. No organism is evolutionarily static. In the case of spaniels, evolution has been dramatically accelerated by selective breeding of "favored" traits by dog breeders.

We may think that animals evolve to reach greater functional complexity higher intelligence and larger size, but there are many cases that counter that notion. Environment plays an important role in deciding which functions are important to have. A fish without eyes is less complex than a sighted one. In the farthest depths of the oceans or in underwater caves where there is so little light that expending energy for eyes is a waste, fish evolved to have no eyes. Their ancestors had eyes and could see, but because their dark environments confer no advantage to having vision, and because it is energetically costly to develop and maintain organs of sight, cave-confined fish whose genes invested less in the development of eyes tended to have a reproductive advantage over those that invested more. Blind fish are less functionally complex than sighted ones, but they have evolved to suit their environment. Thus the blind cave-dwelling fish is less complex yet more highly evolved. This example illustrates one of the central fallacies of the progressivist bias: the assumption that more evolved forms are inherently more complex. If blind cave fish were to suddenly find themselves in a lighted environment, in time their descendents would likely redevelop eyes, because vision is a very useful adaptation. So, while there are trends through time because greater sophistication may confer advantages in the survival sweepstakes, evolution itself is (as it were) as blind as a cave-dwelling fish.

No modern species is directly ancestral to any other. To be so would require that the ancestor ceased to evolve the moment the new species diverged. Modern fishes, for instance, did not stop evolving when some of their representatives began to colonize land. They continue to morph according to the external pressures they are confronted with. The editors of *Fish Cognition and Behavior* attribute the idea that animal life is neatly arranged in a linear progression from fish through reptiles and birds to mammals "to a heady mix of outdated and unscientific thinking."

Without wishing to dignify the idea that being a member of a more highly evolved group makes a species somehow more consequential, you may be surprised to learn that we are—in the genetic sense—not as highly evolved as chimpanzees (though it may be argued that we are more complex). A recent analysis of 14,000 genes found that 233 chimp genes, compared with only 154 human ones, have been changed by natural selection since we shared a common ancestor about six million years ago.[5]

Despite the abundance of such examples, we still use language to refer to other species that betrays our hierarchical views. We call "primitive" those species that have been treading or swimming across planet Earth for a very long time, or those that have relatively small brains. We refer to "lower animals," "lower mammals," and "higher vertebrates." I'm pleased to say that "subhuman" seems to have fallen out of use. When I saw the following title of a paper published in 1973 in the journal *Environmental Research*—"Biological effects of polychlorinated biphenyls and triphenyls on the subhuman primate"—I couldn't help but wonder at the irony that it was the poor victims of enforced chemical exposure who were referred to as "subhuman."[6] Today, the less hierarchical term "nonhuman primate" is used, though we still use them in harmful tests of chemicals, drugs, and military weapons.

The Naked Ape

In Chapter 9, I argued that our perceptions of wild nature are more violent, bloodthirsty, and cruel than wild nature actually is in real life. One of the reasons we feel comfortable with the harsh nature paradigm is that it helps to justify our own transgressions against animals. After all, if nature does it, then we are only taking part in the natural course of things. It also makes it easier to see ourselves as better, more civilized. But are we? How worthy are we of our self-assigned rank as the superior inheritors and rightful landlords of a savage planet?

By any measure, humans are fascinated by violence. The average American child witnesses eight thousand murders and 100,000 other acts of violence on television by the age of twelve.[7] I use this well-worn statistic only to illustrate our cultural preoccupation with mayhem and bloodshed. But these are only staged killings, beatings, and fights. What about the real thing?

Humankind's propensity to "resolve" conflicts through war is well documented. *The Encyclopedia of the Ancient Greek World* lists twenty-three wars fought between the sixth and first centuries, BCE.[8] For less than light reading, web browsers can visit Matthew White's compilation titled *Selected Death Tolls for Wars, Massacres and Atrocities Before the 20th Century.* His list of the "The Twenty Worst Things People Have Done to Each Other" totals just over 380 million humans killed by other humans. There is no overall trend suggesting that we are getting better, for half of these events occurred in the last two hundred years. Top of the list is the Second World War, which claimed 50 to 70 million human victims.[9] According to Colman McCarthy, founder of the Center for Teaching Peace, in Washington, D.C., more than 40,000 people die every month in some thirty-five wars or conflicts worldwide.[10] Figures like these belie claims that we are more civilized and less violent than the wild nature we portray as brutish and cruel.

❖

On a day-long jeep safari I took during a recent visit to South Africa, one of my guides mentioned that if there were hippos in the little Bushman River we were boating down, the boating portion of the tour would have to be canceled because of the risk of being upended by an angry hippo. He added, "hippos kill more people than any other African mammal."

I didn't question his concern about the boat's stability, but his second assertion is plain wrong. While there are no reliable statistics on how many humans get chomped or trampled by angry or scared hippos, it is safe to say that humans kill far more humans in Africa than hippos do. During the hundred-day Rwandan genocide of 1994, up to 800,000 Tutsis were killed by Hutu militia, and as many as 10,000 were killed in a single day. And that is just one of the most notable of dozens of violent conflicts in Africa during the twentieth century, each resulting in tens of thousands, hundreds of thousands, or millions of human deaths. Against numbers like these, every wild hippo alive today would have to be killing several people per year to compete.

I searched online for specific information on numbers of humans killed by wild hippos each year, and although there are several news reports of individuals falling victim to attacks, there were no estimates of total numbers. This in itself is interesting given how widespread are the claims that hippos kill more humans than any other African animal (presumably this figure refers to vertebrates, given that mosquitoes, for instance, are responsible for far more human deaths).

Let's turn the tables. How many hippos are killed by humans? Dr. Rebecca Lewiston, a biologist at San Diego State University and chairperson of the Hippo Specialist Group of the International Union for Conservation of Nature and Natural Resources (IUCN), helped direct me to information on killing of hippos by humans. Exports of hippopotamus tusks in the 1980s and 1990s suggest that between 4,000 and 5,000 hippos were being shot yearly during that period. Between 1991 and 1992, approximately 27,000 kilograms of hippo canine teeth were exported, and in 1997, a single shipment of more than 1,700 hippo teeth en route from Uganda to Hong Kong was seized by customs officials in France.[11] A recent field survey found that hippo populations in the Democratic Republic of Congo have declined more than 95 percent as a result of intense hunting during more than eight years of civil unrest and fighting. Only about 800 hippos remain in Virunga National Park in northeast Congo, down from 29,000 in the mid-1970s, according to Walter Dzeidzic of the World Wildlife Fund.[12] Much of the killing of hippos is for their meat.

There are many other examples we could choose to illustrate the double standard with which we portray animals as dangerous. Wolves, vampire bats, snakes, spiders, and scorpions all come to mind. Sharks are one of humankind's favorite animal villains. Worldwide, fatal shark attacks on humans average less than ten yearly, whereas humans kill between 26 million and 73 million sharks yearly—a ratio of about one to five million.[13] Many die after being tossed back into the ocean having had their fins sliced off to make soup. For sheer savagery, it's no contest: we win.

A few days before my jeep safari I had had a private tour of the South African Institute for Aquatic Biodiversity (SAIAB), in Grahamstown, South Africa, where seventy thousand fishes from about seven thousand species lie preserved in labeled jars. I'm fascinated by fishes, but I find these museums to be morbid places. If you think of a fish as a sentient individual, as something more than a "specimen," then you have to wonder if we ought to be pickling them and stacking them on shelves like this. About twenty years ago, American ichthyologist Mike Howell of Samford University in Birmingham, Alabama, began wondering the same thing. After years of mounting discomfort with collecting fishes and putting them in formalin-filled glass jars, most of which were never opened again, Howell decided there must be a better way to catalogue his fishes. He came up with a simple, ingenious solution: a narrow, V-shaped, glass-sided tank that could be taken into the field. When a small fish is caught, it is gently transferred with some native water into the tank. Here, the restrained fish can continue to breathe, and one can take detailed, close-up photographs. Thus, a catalog can be built without killing fishes. These photographs show the fishes in their full living color (dead and pickled fishes lose this), and can be used to identify most fishes. Howell has used his T-P (teaching-photographic) tank with hundreds of students and estimates that this method has saved thousands of fishes. Howell informs me that about a hundred of the tanks have been selling in an average year since he received the patent on it in 1993, and that it is slowly but surely making its way into the educational system.

At SAIAB I asked the curator, Willem Coetzer, why it was deemed necessary to kill all these fishes to identify them. He replied that keying a fish to species sometimes requires attention to minute details, such as counting fin rays and scale ratios. I knew this from my undergraduate biology days, when I keyed out many fish in an ichthyology

Ichthyologist and inventor Mike Howell (in cap) views a fish in one of the teaching/photographic tanks he developed, which has spared many thousands of fishes from being preserved in glass jars. (Photo courtesy of Mike Howell.)

course. I asked if he had heard of the T-P tank. He had not. I sent him information on the tank in hopes that some fishes might be spared in the future.

For all of our capacity for civil behavior, we humans are also capable of stooping to the depths of uncivility. Humans are the perpetrators of the most depraved behavior toward animals. Whalers often harpooned babies first so they could then strike at adult pod members who returned seeking to aid their dying comrade.[14] When overhunting by European settlers sent deer numbers diving, wolves were blamed. In retaliation, most were shot, clubbed, poisoned, or trapped; some were tortured using unspeakable methods. Today, one can watch online videos of factory farms, fur ranches, abattoirs, seal hunts, puppy mills (large-scale commercial pure-bred dog breeding facilities often operating under inhumane conditions), and other operations that show no less savagery than in earlier times. Some of what we do is the result of deliberate cruelty; most is the result of greed and institutionalized indifference.[15] Recently I was asked to provide expert comments on seven minutes of undercover video filmed at an American dairy farm. What I saw was far from the worst I've seen in twenty years of working for animals, but still it pained me to watch (among other abuses) calves struggling and gurgling in pain as their horn buds were removed with searing hot pliers and their tails were severed without painkillers.

In sum, instead of giving animals the benefit of the doubt in questions of animal intelligence and feelings, we continue to give the benefit *to* the doubt. Rather than put the burden of proof on those who would deny that a cat is conscious or a porpoise feels pain, traditional science from the time of Descartes in the seventeenth century until late in the twentieth has asked "How can we know?" and demanded proof from those who dare to affirm the animals' case.

Biology and Biography

Finding a better relationship with animals depends, in part, on our seeing them as autonomous individuals with feelings and lives worth living. In the current way of things, animals are too often categorized only into groups of *things,* such as species or populations, or renewable resources. Species and populations are useful concepts in many fields such as biology, ecology, and genetics. But these terms fail to fully define animals because they are not sensitive to the experiences of sentient individuals. When we refer to mountain lion "harvest limits," "whaling quotas," or "fish stocks," individuals cease to exist. Regarded in those terms, we may as well be talking about tons of sand.

Viewing animals as species or individuals can lead to disparate conclusions. Depending on which perspective you take, the domestic chicken, *Gallus gallus,* is the most, or the least, successful species of bird on our planet. From a purely species-centric perspective, chickens are supremely successful because they are now so abundant. Some ten billion live and breathe each year in the United States alone. Those selfish chicken genes have had a splendidly good go of it, building new chickens at a rate that would have been the envy of even the passenger pigeon before it was made extinct.

But such a view disregards the sentient individual and sees only the nonsentient species. Viewed from an individual's perspective, the chicken is the least successful bird on earth, because most of them are made to suffer on factory farms and in slaughterhouses.

When a species becomes endangered (usually due to human causes), we readily sacrifice the well-being of other animals who may be interfering with the endangered species. Heaven forbid that coyotes should dilute the genetic purity of red wolves by interbreeding with them—red wolves being a species that humans practically drove to extinction in eastern North America. According to species conservation doctrine, genetically pure wolves must be maintained. The United States Fish and Wildlife Service is sterilizing coyotes that live near the wolf populations to try to prevent hybridization. Notably, it is still legal for cattle ranchers to kill red wolves if they threaten the source of someone's hamburger. In Europe, the Royal Society for the Protection of Birds has been killing ruddy ducks—a North American species introduced by humans to enhance captive wildfowl collection—for the offense of interbreeding with its close relative, the endangered white-headed duck, thereby tainting the genetic purity of the latter. The white-headed duck's rarity is due to hunting and habit loss. It is a common theme of ours: we persecute the members of one species—in this case the white-headed duck—to the point of rarity, then we turn our sights on another, the ruddy duck, for its marginal impact on the first species. We may sympathize with the efforts to secure the protection of endangered species, but doing so at the expense of other animals is misguided, and hypocritical when we continue to threaten the endangered species through our own activities.

One way to illustrate our moral hypocrisy toward members of other species is to substitute the human for the animal in our discourse. Here's an example from a recent article concerning a decision to resume the shooting of elephants in South Africa following a thirteen-year moratorium. I use **boldface** to indicate where I have replaced the word "elephant/s":

> South Africa will allow **people** to be killed in an attempt to control a burgeoning population, the government said, setting a trend that could embolden other southern African nations confronting the same dilemma. As outraged **human** rights activists threaten to promote tourist boycotts, Environment Minister Marthinus van Schalkwyk said Monday the government was left with no choice but to introduce culling "as a last option and under very strict conditions." There would

be no "wholesale slaughter," he promised. "Our simple reality is that **human** population density has risen so much in some southern African countries that there is concern about impacts on the landscape, the viability of other species and the livelihoods and safety of **wildlife [people]** living within the **humans'** ranges," he said. . . . "They are doing the responsible thing," said Rob Little, acting chief executive of WWF South Africa. . . . "We all love **people**, no one wants to kill them, but we don't have the luxury for one species to dominate," he said.[16]

The irony of that doctored paragraph is that *human* population density has indeed risen, it does have negative impacts on the landscape, and the livelihoods and safety of wildlife are profoundly affected by our geographic dominance. But the original article is about the elephant problem, not the human one. There are many times fewer elephants in Africa today than there were in historical times, and human encroachment and persecution are the chief causes of their declines. Of course I am not advocating culls of the human population, but I am pointing out the hypocrisy of making elephants pay for our own lack of restraint.

Wildlife management has a long history of meddling at the behest of special interests or simply through prejudice. In East Africa, wild dogs were shot on sight because their methods of killing were considered "unethical." Game wardens in Kruger National Park killed all predators as a matter of policy that persisted into the latter twentieth century. Early settlers and hunters in Africa considered lions as vermin and shot them on sight. Because of shooting and poisoning, hyenas have been eradicated from much of their former range.[17] The same goes for mountain lions, wolves, and bears in North America. The only way we can rationalize their elimination is through the delusion of our supremacy.

One of the advantages of an individual-based perspective of animals as opposed to a species-based view is that every good deed is palpable. The anthropologist Loren Eiseley once encountered a young man on a beach strewn with millions of stranded starfish dying in the sun. As the young man tossed another starfish into the surf, Eiseley politely informed him that there are miles and miles of beach and that his actions could not possibly make a difference. At this, the young man bent down, picked up yet another starfish, and as it met the water, said: "It made a difference for that one."[18] The Talmud expresses this principle thus: "Whoever destroys a single life is as guilty as though he had destroyed the entire world; and whoever rescues a single life earns as much merit as though he had rescued the entire world."[19]

Similarly, with tens of billions of chickens being consumed yearly by humans, making a difference might seem overwhelming. But chickens are individuals, so helping just one is a measurably good deed. When many people do their part, the numbers soon add up. In April 2005, a forty-five-year-old British woman named Jane Howarth started an organization called the Battery Hen Welfare Trust, which rescues and resettles chickens as an alternative to slaughter. As of June 2008, the organization had grown to a team of eighteen rescue coordinators and forty-five volunteers across Britain. Howarth estimates the trust has found homes for 92,000 chickens.[20] In the United States, Karen Davis left her job as an English professor in 1990 to form United Poultry Concerns and to write several books dedicated to the protection of chickens and turkeys.

Ignoble Science

We think of vivisection as cutting open animals while they are alive or conducting chemical tests on them, but it can more generally be defined as causing deliberate harm to an animal in the name of science.

Vivisection is a diverse enterprise. A 2008 statistical analysis conducted jointly by two British animal protection organizations—the Dr. Hadwen Trust, and the British Union for the Abolition of Vivisection (BUAV)—concluded that about 115 million animals are currently used each year in laboratory research and testing worldwide.[21] Funding for medical research rose from $37 billion yearly in 1994 to $94 billion in 2003.[22]

My first encounter with vivisection was as an undergraduate student. I took a wrong turn in the biology building of my college and found myself wandering the basement corridors. This was in the days before security was tightened for fears of break-ins by animal rights activists. As I passed an open door I paused to stare at several white rabbits strapped onto their backs. Resting on each of their bellies was an inverted container wrapped in tin foil. The rabbits' eyes were rolled up into the backs of their heads. I learned later that the rabbits were being used to feed blood-sucking reduviid bugs for parasite research.

Not all vivisection is done in indoor laboratories. A few years later, I sat in the audience at a biology conference as a graduate student described her field studies on the reproductive ecology of wild mourning doves. Her methods shocked me. Her study design required obtaining the doves' brains. On locating a nest and climbing to it, she would systematically shear off the heads of every chick with a pair of sharpened scissors. I wondered how she could steel herself to commit such a grisly act.

Toxicity testing on animals is perhaps the plainest manifestation of humankind's ability to hold ourselves on a higher plane of righteousness. Here's an example of just one of hundreds of such protocols going on at any one time in labs throughout the world. Four hundred rats and four hundred mice were exposed to abnormally high doses of the test substance citral over various durations for up to two years to assess its potential to be toxic or cancer-causing (carcinogenic) to humans. Citral is a lemon flavoring commonly added to processed foods, drinks, and candies. The animals fared poorly. In the second week of the fourteen-week trial, all of the rats and some mice in the highest-dose group were already approaching death, with symptoms including listlessness, hunched posture, absent or slow paw reflex, dull eyes, emaciation, lethargy, and various forms of internal tissue and organ damage.

When I read research protocols like these, I find myself shaking my head in wonderment that my species can methodically poison sentient animals, observing with clinical detachment and taking meticulous notes and measurements as they slowly die. It is terrible when someone lashes out violently and harms or kills an animal in an act of passion. But there is something uniquely chilling about the dispassionate way that animals are killed in toxicity tests. George Bernard Shaw, a staunch opponent of vivisection, said, "custom will reconcile people to any atrocity."

Here's a little vivisection puzzle for you. See if you can spot a small change I've made to the following sentence:

> In new research that may advance the search for treatments of human depression, a protein that seems to be pivotal in lifting depression in moose has been discovered by [a Nobel Laureate researcher from a midwestern American university].[23]

Bravo if you guessed that moose have not been used in depression studies and that the original word was actually "mouse." The mouse is the mammal of choice for laboratory studies aimed at improving human health. Credible sources place the number of mice used in vivisection today at around one hundred million per year.

You earn bonus points if you recognized a deeper problem with the statement above. It is the implicit assumption that a "mouse model" of depression is a useful predictor of human depression. There are several reasons to doubt the efficacy of such a model. First of all, mice are not humans. For one thing, they absorb and metabolize drugs differently; the secretary of the U.S. Department of Health and Human Services confirms that 92 percent of drugs found to be safe and effective in animal tests are unsafe or ineffective in humans. Mice in labs also live much shorter lives, they run on all fours, communicate by means of urine marks (see Chapter 2), and—especially disadvantageous in studies of depression—they cannot report to us with verbal language how they are feeling.

Some of the mice used in this research are not normal mice to begin with. Many are transgenic "knockout" mice lacking a protein, p11, that appears to reduce depression symptoms in humans, and what are presumed to be depression symptoms in mice. These mice are called "helpless" because when they are dangled by the tail for six minutes using adhesive tape (an often-used method called the "tail suspension test"), they don't try to right themselves as much as do mice with the p11 protein. This lowered willingness to resist being hung by the tail is taken as an indicator that a mouse is depressed.

For me, though, what is most deeply problematic with this sort of research isn't about science. It's about ethics. These experiments also entailed daily injections for up to four weeks, electroconvulsive treatment delivered by ear-clip electrodes, and the tail suspension test mentioned above. Killing methods involved decapitation under anesthesia and focused microwave irradiation, in which a powerful microwave beam is aimed at the head of the restrained animal. Why is it acceptable to treat sentient animals in ways that we would never condone for humans? What does a mouse lack that makes it conscionable for us to subject him to harmful experiments that we would describe as sadistic or worse if done to a human?

The mice aren't the only ones to suffer for these studies. The depression research team used fetal bovine serum, a widely available laboratory serum used to nourish the growth of cell cultures. It is obtained at the slaughterhouse by inserting a large needle into the heart of an unanesthetized fetal calf who is at least three months old and removed from a freshly slaughtered pregnant cow at the abattoir. The research team also used rabbit polyclonal antibodies whose production involves repeatedly injecting rabbits with irritants to trigger a massive immune response, then drawing out the fluids with a syringe.

I did not choose this example because it was unusual. The researcher is just one of the more highly decorated among legions of skilled scientists who earn their daily bread by animal research. The practice of vivisection subscribes to a view that growing numbers of humans question on ethical grounds. It is the view that we are morally entitled to inflict severe harm on other sentient animals in the pursuit of our own interests—sometimes no more critical than lemon flavoring—which we conveniently designate as more compelling than theirs. It is, in short, the idea that we have the right to do something because we have the power to do it. One of the plainest moral objections to this might-makes-right attitude is that it is vulnerable to the "intelligent alien" scenario: If the might-makes-right justification were valid, then we must concede that there would be nothing immoral about a "superior" race of aliens arriving to enslave, kill, and eat us.

Moral Toddlers

For all our moral sophistication in theory, in practice we remain moral toddlers. Our colonial history has been a process of lack of restraint—take what you have the power to get. European conquests and colonialism, and slavery in general, epitomize a might-makes-right, take-all-you-can attitude in much of human history over the past thousand years, and one that continues to characterize humans' current relationship to nonhumans.

Colonialism was aided and abetted by our capacity to exploit animals. In his Pulitzer Prize–winning book *Guns, Germs, and Steel: The Fates of Human Societies,* Jared Diamond explains that disparities in the suitability of local mammals for domestication—as sources of food, skins, and transport of humans and goods—between Europe on the one hand and Africa and the Americas on the other played a key role in the former's conquest of the latter continents. In North America these imbalances were exacerbated by the extinctions of most of the former large mammal species. Furthermore, humans caused or at least contributed significantly to these extinctions.[24] So it can fairly be said that using and killing animals is interwoven with the history of interhuman conquest and enslavement.

Today, traditional colonialism is largely over, with virtually all former colonies having gained their independence during the past century. This change is a sign of moral progress, of the decline of might-makes-right as an acceptable way to conduct our affairs.

But when it comes to animals, our moral flower is still waiting to bloom. Let's examine an example pertaining to the pursuit of bullfighting.

A morning radio story on bullfighting included this comment by a female fan: "I'm not too concerned for the bulls; they're bred for this purpose." This is a variation on a familiar theme you may have heard applied to animals raised for meat, animals bred for laboratory experiments, and fish "stocked" in lakes.[25]

The problem with the "bred for that purpose" argument is that it has no bearing on the moral object—the suffering of the animal. If being bred for the bullfight somehow meant that the animal was insensate to pain and suffering, then that could have some bearing on the rightness or wrongness of the act. But there is no evidence for this. A bull destined for the bullring, a mouse caged for a carcinogenicity study, and a salmon fattened for the angler's hook are no less sentient than those more fortunate individuals with a life of freedom.

Another commenter responded that "while it [bullfighting] may be brutal... it isn't gratuitously torturous," and "I suppose it's a bit like any other sport: if you're not a fan then don't watch." This is the "out-of-sight, out-of-mind" argument. In sophistication it rivals the infant's failure to recognize the permanence of an object that has just been placed behind a screen. A sporting event that causes a lingering, violent, and painful death is unavoidably gratuitous. Suggesting that those who are not fans should simply not watch is like advising bystanders to look away rather than act as a crime is being committed.

Most of us actively participate daily in the moral fiasco of out-of-sight, out-of-mind. Whenever we go to the supermarket and purchase meat or dairy products, we support some of the cruelest treatment of animals that humans currently perpetrate. That we don't ourselves perform the various insults and deprivations of factory farming does not lessen the suffering of the animals so treated. If we buy it, we support it. By inserting the middlemen, we are conveniently sheltered from the atrocities.

If there is one quality in humans that forms the foundation of our dominion over other animals, it is our intellect. The chief justification for our dominion over them is that, compared to us, they are "dumb brutes." We like to point out apparently mindless behavior in animals. This can take the form of erroneous tales passed on through generations. That turkeys drown from gaping skyward during storms is one such falsehood. Another way to dismiss animals is to point to behavior that seems mechanical and stupid. Turkeys are commonly dismissed as intellectual lightweights because they will attempt to mate with clearly inanimate models. (I have two words for those who assume this proves stupidity: inflatable doll.) But more to the point, is their intelligence really relevant to their moral standing?

Is it really morally right to discriminate against other beings because they are not as smart as us? We do not apply such thinking to our own kind. At the same time that we deny any rights to perfectly healthy animals, we afford full rights and privileges (to the extent that they can be exercised) to the mentally handicapped.

Animals' brains tend to be specialized to meet their ecological needs. Most bats who use echolocation have little brain tissue assigned to processing visual information, but their auditory processors are comparatively enormous. Similarly, about one-third of the naked mole rat's brain is devoted to processing information from the area around the teeth, compared to about one percent in a human. The mole rat digs burrows, manipulates tools, and performs many other tasks with its protruding incisors, which can be spread apart and moved independently like chopsticks. I imagine a walrus has a lot of neurons dedicated to those sensitive whiskers described in Chapter 2. Animals are good at what they do. As Albert Camus said, "In the water, the turtle becomes a bird."[26]

It has never seemed right to me to use an animal's perceived intelligence as a yardstick for how we may treat it. For one thing, it is not easy to appreciate the intellect of another species, whose mental skills may not be apparent to us, especially in the artificial conditions of captivity, where there may be few opportunities to perform natural behaviors. No animal, human or nonhuman, has any choice in the brains or raw intelligence they are endowed with. Evolution does not blindly favor bigger or smarter brains.

Intelligence isn't absolute. An animal (including a human) may be highly intelligent in some ways and dull in others. There are various ways in which lifestyle characteristics can influence cognitive skill. Ravens, for example, are proficient at choosing which of two crossed strings need to be pulled to hoist up food, as Bernd Heinrich explains in *Mind of the Raven,* whereas dogs tend to pull on the wrong string.[27] Dogs, however, are remarkably good at interpreting signals from people. They'll understand nuanced gestures in humans such as gazing or pointing as a means of locating food. They intuitively perform better at following the gaze of another—known as mutual looking—than do great apes.[28] As we saw in Chapter 4, chimpanzees far exceed humans on short-term spatial memory tests. Animals are good at performing mental tasks that are evolutionarily relevant to their species. Humans have gone to the moon, but we don't have a dog's ability to sniff out explosives or cancer.[29]

Brains are expensive organs to grow, maintain and operate, and there are good reasons for many creatures to keep their brains small. It is illogical to abuse and "punish" those with lesser brains, because they may be no less sensitive to pains (or pleasures) than smarter individuals; to discriminate on this basis is no more sophisticated than on the basis of the length of the legs or the number of vertebrae. Intellicentric thinking skirts the bigger issue of sentience. If animals can think, have emotions, and can suffer, then they are entitled to moral consideration because their lives have value independent of any utilitarian value it might have to "superior beings."

The humans-are-superior argument is also vulnerable to the Argument from Marginal Cases. This argument, forwarded by the prominent Princeton University philosopher Peter Singer, author of the seminal book *Animal Liberation* (1975), holds that if animals have no direct moral status due to their lack of, say, rationality, then neither do other members of human society such as the senile, the comatose, and the mentally enfeebled. Yet few of us would permit the use of these humans in harmful experiments.

For all our philosophical and moral sophistication, in practice we're not very sophisticated when it comes to trying to justify institutionalized cruelties to others. We can't have it both ways: if we are on a higher plane than other animals, then we should know better than to treat them as means to our own ends, as though their sentience, unlike ours, were of no consequence.

The Extinction Folly

The irony of our unique intelligence is that it continues to get us into trouble. We fail to foresee, and are slow to react to, problems of our own making, such as global warming, or the extinction of species. I remember the sadness I felt when, as a boy living in New Zealand, I learned that between thirteen and twenty-five species of giant flightless birds known as moas, ranging from about the size of a turkey to that of an ostrich or larger, had roamed the island before being driven to extinction by Maoris just a few centuries earlier. As I gazed up at a reconstructed skeleton at the museum, I wondered how exciting it would have been to have seen these huge birds in the flesh.

Species extinctions are an ongoing, natural process, but every once in a while, apocalyptic geological upheavals such as volcanic eruptions and extraterrestrial objects slamming into the Earth cause a major spasm of extinctions over a relatively short time-span (a few hundreds or thousands of years). The fossil record indicates that there have been five episodes of mass extinction during the previous 400 million years. The sixth and current mass extinction is the first that can be attributed to a single species. As far back as 1993, Harvard biologist Edward O. Wilson estimated that 30,000 species are going extinct on earth each year, which equates to three species per hour.[30] Most of these would be tiny, obscure creatures not yet documented by scientists. Baseline extinction rates are difficult to estimate, but most experts agree that current extinction rates are abnormally high, and we are the main cause.

It is likely that if the Maori hadn't driven moas to oblivion, European settlers would have made short work of finishing the job, as they did with the Carolina lorikeet and the Labrador duck in North America. On Mauritius, the dodo was clubbed and cooked to extinction within a hundred years of the arrival of Portuguese sailors and Dutch settlers.[31]

We are no less efficient when we turn our sights to the creatures of the seas. In his 1984 book *Sea of Slaughter,* Canadian biologist and novelist Farley Mowat recounts the destruction of whales, seals, shorebirds, seabirds, and fishes, mostly for insatiable European markets. Faster than a fish in the water, the great auk was clumsy on land, and they had no defense against the paddle-wielding hunters who came for their eggs. At the height of the egg trade, teams of men swept the auks' rocky island rookeries in a sequence, crushing all the eggs as they went, then returned a week later to collect all of the newly laid eggs to ensure they were fresh. By 1802, the last great auk had fallen. Thirty-four years earlier, the Steller's sea cow, the walrus's larger cousin, had been hunted to extinction for its skin, its meat, and its fat.[32]

In his book *Collapse: How Societies Choose to Fail or Succeed* (2005), Jared Diamond lists seven commercial fisheries that have collapsed due to *overexploitation* (a dubious concept if you think exploitation is wrong to begin with).[33] Before 1900, total global annual fish catch had never exceeded 10 million metric tons. A million metric tons is about 2.2 million pounds, and probably represents a similar number of individual fishes. By the 1950s it had doubled to 20 million tons (about 4,400,000,000 fishes), by 1960 it had doubled again to 40 million tons (8,800,000,000 fishes), and doubled again to 80 million tons (16,600,000,000 fishes) by the 1970s. In the 1980s, the oceans refused to give up any more fishes, and population collapses have ensued.[34]

It should be added that these are official statistics. Pirate fishing is estimated to value between $4.9 and $9.5 billion yearly. Then we must factor in the ugly reality of *bycatch,* a term that refers to unwanted fishes and other animals unwittingly ensnared in the nets or snagged by hooks. Global average bycatch rates are between one-quarter and two-thirds of the desired catch. Doubling the above numbers might give us a more realistic estimate of the toll.

Extinctions, or near-extinctions, have secondary or "collateral damage" effects for other species that depend in part for their livelihood on the afflicted species. When whales die a natural death they sink to the bottom of the ocean, where they become important islands of food for benthic ocean-dwelling scavengers (*benthos* is Greek for the depths of the sea). No one knows what impact the near extinction of whales has had on these communities. The annual number of whale falls, before commercial whaling began, is likely to have been hundreds or perhaps thousands of times higher than today, according to Rachel Cave of the National University of Ireland.[35] It is one of the many ways that our actions may profoundly affect unknown ecosystems without our giving any thought to it.

Environmental writer Terry Glavin estimates that today, a species goes extinct every ten minutes. Currently, about one in eight bird species, one in four mammals, one in three amphibians, one in four turtles, half of 1,700 fish species surveyed (out of 28,000 total species), and about one in eight plants is threatened with extinction.[36] Elimination (e.g., hunting) and habitat loss are the leading causes of extinctions, but as our collective ecological footprint expands, we are coming up with new ways to put an end to species. In a 2004 study published in *Nature,* a team of nineteen scientists estimated that climate change may drive a quarter of land animals and plants extinct.[37]The United Nations Intergovernmental Panel on Climate Change (IPCC) estimates that up to 30 percent of species are at risk of extinction.[38] Species disappearances from global warming are not hypothetical; they are already being recorded. Mountain-restricted species are among the most vulnerable, because they have nowhere else to go to seek cooler climes. Two of every three species of harlequin frogs have disappeared over the past twenty to thirty years in the tropics of the Western hemisphere.[39] Climatic changes are also thought to be more favorable to the chytrid fungus that is now plaguing many frog species worldwide.[40] These are only a few of the countless examples of climate change's ill effects on animal life worldwide—including many species we may never know about.

It has been said that every extinction brings us closer to our own.

CHAPTER ELEVEN

the new humanity

"How wonderful it is that nobody need wait a single moment before starting to improve the world."

—Anne Frank

I once found a flicker lying prostrate next to a large glass window in rural Ontario. These robust woodpeckers range throughout North America, and their distinctive "wicka-wicka-wicka" calls are familiar to anyone who attends to woodland sounds. The bird was clearly not dead and I gently picked him up. As he lay there cradled in my hand, I felt fortunate to see nature's graphic design up close on a bird I'd only ever enjoyed watching from a distance. The white breast had a spray of bold black spots beneath a crescent-shaped black bib. A red crescent graced the blue-gray nape, and a black bar extended from the mouth across each tan cheek. The undersides of the wing and tail feathers were buttered with bright yellow. The Austrian painter Gustav Klimt might have drawn inspiration from these birds.

I have found about two dozen birds knocked out this way after flying into windows. One can't be sure the bird will recover. From my limited experience, birds who don't break their necks on initial impact usually pull through and recuperate fully. The flicker revived, and it was exhilarating to see him rally, collect his strength, and surge from my hand.

That bird is an allegory for our relationship to other animals. The window—mirroring an alluring grove of shade trees—reflects the kindness we may show animals, and its cold, hard surface is our institutionalized indifference. Although our relationship with animals is by some measures at its lowest point in history, the brighter news is that change is imminent. Our imperialist view of animals is being rocked by our growing awareness of their capabilities and their sentience, and the inescapable fact that on an interdependent planet what befalls them befalls us. I believe that like the flicker, our relationship to animals is poised for flight.

The End of Growth

Human population growth—and the concomitant increase in human consumption of resources—underlies some of the most serious problems faced by animals, including humans. Conventional wisdom holds that the planet is filled to capacity, and there isn't room for any more consumers, human or nonhuman. When we add more, others have to make room. Which means that as the human population grows, other organisms are inevitably being pushed out. The rising extinction rates discussed in the previous chapter illustrate this.

Of all the academic disciplines, economics is perhaps the most influentially at odds with the problems of growth and human overpopulation. In the current scheme, growth is integral to economics, which seeks to build trade and aid development. The goal of every commercial enterprise is to sell more. To do that, either there must be more consumers, or consumers must consume more per capita. In the current version of capitalism we are bombarded with the message that growth is good. Companies are proud to associate themselves with growth. The day I wrote this, a bank ran a full-page advertisement in the *Washington Post* proclaiming itself in big bold letters to be "A BIG FAN OF GROWTH."

Most of us have heard the phrase "sustainable growth," yet few know what it refers to. As authors Herman Daly and Kenneth Townsend point out in their 1993 book *Valuing the Earth: Economics, Ecology, Ethics,* sustainable growth is both "an economic oxymoron" and "an impossibility theorem."[1] In a finite system (exhibit A: planet Earth), sustainable growth is as unattainable as perpetual motion; it violates basic laws of physics. It has been calculated that if twentieth-century rates of human population increase continued for the next thousand years, a mass of humanity would cover the earth shoulder to shoulder more than a million deep; another thousand years on and the mountain of humanity would be approaching the edge of the known universe, traveling outward at the speed of light.[2]

This imaginary scenario illustrates the inevitable link between economic growth and ecological sustainability. It also shows that any link between growth and a higher quality of (human) life is tenuous, and at best temporary. As long as economic models are defined by growth in consumption, ecosystems will increasingly feel the strain as more human consumers populate the land, and as more resources get used up.

I sought out the opinions of three influential economists, whom I chose at random from a Google search of economics departments. I emailed them the following question:

> It has always been my understanding that the economic paradigm is aimed at sustaining (economic) growth and development. How do we reconcile the concept of economic growth with the fact that sustained ecological growth is unattainable in a finite system?

All three replied to me the same day. A professor of economics at George Mason University, recommended the works of the late Julian Simon, once an economist at the University of Maryland and author of *The Ultimate Resource: People, Materials, and Environment* (1981). I consulted Simon's book, and found it engaging and scary. In a chapter on extinction, Simon wrote that "recent scientific and technical advances—especially seed banks and genetic engineering—have diminished the economic importance of maintaining species in their natural habitat." Such a cavalier attitude to the integrity of natural ecosystems is anathema to ecologists and biologists. Simon also saw knowledge as the main constraint on human growth, and because the source of knowledge is the human mind, he believed that having more humans on earth was good because it increased our knowledge pool. This is a circular argument: we need more people to create innovative responses to the challenges presented by having more people. By that logic, we need more global warming to stimulate solutions to global warming.

A professor of economics at Princeton University expressed a similar sort of view in an e-mail. The gist of his reply was that productivity shows steady increases through innovation. That is to say, technological advances allow us to avoid the limitations of living in a finite system by more efficient production.

Philippe Aghion, a professor of economics at Harvard University and coauthor (with Peter Howitt) of *The Economics of Growth* (2009), invited me to phone him. Here is a brief excerpt from our conversation.

Jonathan Balcombe: Is growth fundamental to the field of economics?

Philippe Aghion: Before, we had sustainable growth. Now the challenge is to make growth sustainable with environmental constraints.

J.B.: But surely at some point, on a finite earth, we have to stop growth in consumption?

P.A.: Not necessarily. . . . At some point we may not care that we don't have oil because we can rely on solar and other energy sources. . . . By directing research toward innovation, we can make up for resource limitations. Growth is driven by innovation.

J.B.: Okay, but at some point things must level off. Surely, you don't believe the earth could support, say, 100 billion humans?

P.A.: Accelerations in human population during the past century have been due to the Green Revolution [technological advances in agriculture] and the health revolution [advances in health care]. But now we are realizing that with development comes also a trend toward controlling fertility. Children are living longer, which in turn is driving a deceleration in fertility rates. I expect we will see a deceleration in human population growth happening in the next thirty to forty years. It is another reason to be less pessimistic.

J.B.: When we speak of "population," we are usually being anthropocentric. We aren't the only occupants of earth. We share the planet with albatrosses, lions, and codfish and a million other species. To the extent their population declines can be connected to human activities, is that not also an argument for reigning in human population growth?

P.A.: We should be concerned about these other populations, and we should have policies to benefit them. I am totally in favor of pro-environment policies. I favor a carbon tax, and nations should make their international trade contingent on their trading partners enforcing sound environmental policies.

J.B.: What about people's individual responsibility? For instance, I am a vegetarian for humanitarian and environmental reasons. Do you and other economists advocate that individuals take personal responsibility?

P.A.: Economists don't do that sort of thing, but we should. I'm not a vegetarian myself and I feel very guilty about that, but I watch how much I eat. I am definitely in favor of personal responsibility.

J.B.: Any final remarks?

P.A.: We need awareness, innovation, and action. But I see no reason to have a Malthusian view. A combination of the two forces—fertility deceleration, and innovation—are keys to future sustainability. It is crucial to induce contraception as much as we can, everywhere. The ultraconservative sort of thinking that opposes such measures is extremely damaging.

In sum, my sense is that Dr. Aghion's position represents a sort of constrained optimism. On the one hand, he seems quite confident that through innovation we will meet the challenges created by our growth. But that doesn't mean we should passively react to crises as they arise. Aghion advocates proactively curbing our growth through wise public policy and education.

My brief foray into the world of economics left me with the feeling that we are still largely in denial over the grave consequences of our numbers and their impact on the planet and its other denizens. There's no question that innovation has mostly met the needs of our growing numbers since Thomas Malthus famously predicted in 1798 that humankind would suffer massive die-offs as the finite food supply would fail to meet the demands of a geometric growth in human numbers.

But at what price our expansion? We can be certain that with more humans there will be fewer animals living free in the world. Animals need wild places to live, and we continue to take them away. Nearly half of all tropical rain forests worldwide have been destroyed for human use, and about one percent of what remains is being taken away each year.[3] Other habitats are declining, too, including wetlands, grasslands, temperate deciduous and tropical dry forests. In terms of pure numbers, we have reached the nadir of our relationship with animals because—as the editors of a 2006 volume titled *Killing Animals* point out, we kill more of them today than ever in our history. We are also sending them to the oblivion of extinction at higher rates than before. Anyone keeping a score sheet could justifiably conclude that we are getting worse.

I remember visiting the only marine estuary habitat left on the island of Crete in 1998. This precious jewel of coastal wetland turned out to be a patch of land not much larger than a typical suburban driveway. Several rare shorebird species probed for invertebrates in the muddy sand. Ominously, a bulldozer was leveling the adjoining plot immediately inland. As I watched some of the bulldozed clods roll down onto the estuary, I realized that the machine could easily have buried the jewel in less than an hour.

Having more humans on earth does not improve the quality of life for humans, either. Human overpopulation has strong links to poverty and hunger (admittedly not new problems), pollution and climate change. In tropical regions, local population density has been directly correlated to the poverty status of the local people, most of whom lack an education in family planning.[4] Human overpopulation is driving climate change through loss of trees and the burning of fossil fuels.

These ills denote a self-centered ethic—an unwillingness to restrain ourselves. A paradigm shift in humanity's relationship to the planet and its other life forms requires the acknowledgment that *growth* is no longer a good thing. Chief among those things that need to stop growing is the human population.

Today, however, addressing human overpopulation remains firmly off the public policy agenda. This is paradoxical when the problem either fosters or exacerbates so many of the challenges faced by modern societies: hunger, gridlock, habitat and biodiversity loss, water shortages, violent conflict. The idea that growth is progress is an anachronism that today serves only those relatively few who profit from another residential development and a longer line at the cash register.

The question is not whether or not we should abandon the notion of continued population growth, but how we want to deal with the certainty that growth will end (recall the scenario of humans standing shoulder to shoulder across the planet's surface). Meanwhile, as long as we continue to allow our numbers to grow, we will witness, among other things, the continued shrinkage of natural habitats and the organisms that live there. A world with fewer people—sharing the responsibility to reproduce below replacement—will be better for the planet, for animals, and for us. There is nothing misanthropic about that. With intended levity, the poet Ogden Nash once said, "Progress might have been alright once, but it has gone on far too long." The same goes for growth.

The Trouble with Meat

As human populations continue to expand, so do the number of mouths that need feeding. Increasingly, what is going into the mouths of humans is animals. It is here that the connection between human growth and animal well-being is most direct. Our growing meat habit represents, arguably, the greatest direct threat we pose to the welfare of animals. And to ourselves.

In June 2008, *New Scientist* magazine invited readers to submit their votes for the world's deadliest weapon created by humans. I believe the hands-down winner is the table fork. Each year, this humble utensil abets the deaths of millions of humans by conveying into their bodies steaks, chicken breasts, and other foodstuffs known to cause heart attacks, cancers, strokes, diabetes, and other diseases. According to the World Health Organization, approximately 17.5 million people died of cardiovascular disease alone in 2005. This is nearly a third of all human deaths globally.[5]

Since most of these harmful foods are of animal origin, and since the magazine didn't specify *human* lives claimed, we might add to the yearly death toll the tens of billions of animals killed to be eaten with forks. To the fork we should probably add the chopstick. A study by the Pew Commission found that in three decades meat consumption in China rose from 44 to 110 pounds a year. China is now the world's most carnivorous nation, accounting for one-quarter of the world's meat consumption.[6]

Few people yet realize that what they put on their plates has more environmental impact than how much they fly or drive. Animal husbandry takes up more land than any other activity by humans, and it accounts for 70 percent of all human consumption of fresh water. Meat production causes more greenhouse gases to be released into the atmosphere than does all global human transportation combined—18 percent compared to 13.5 percent, according to recent estimates by the Food and Agriculture Organization (FAO). The FAO also estimates that animal agriculture is responsible for over two-thirds of all human-caused nitrous oxide and methane emissions, respectively. As climate-warming compounds, these gases are 265 times and 23 times as potent as carbon dioxide.[7]

One may reasonably ask why raising animals is so environmentally destructive. It is not as though eating meat is unnatural. Many animals evolved to do so, and historically humans are omnivorous. There are two main reasons for the trouble with meat. The first is the ecological principle discussed in Chapter 9 (see the section titled "Peddling Predation"). Recall that food chains operate only at about 2 percent efficiency. The remaining 98 percent of available energy is lost (as heat and waste, for instance) at each level of a food chain. Thus the biomass of a tier of predators is many times lower than the biomass of herbivores on which it feeds. Second, humans haven't evolved as carnivores, and the planet is ill-prepared to accommodate seven billion large "naked apes" with meat squarely on the menu. Feeding meat to billions of humans is unsustainable.

The flow of energy through food chains also helps to resolve a common misconception—that we must kill more plants to sustain someone on a vegetarian diet than a meat-eater. The opposite is true. Meat-eating—by consuming herbivores—accounts for the destruction of far more plants than does a plant-based diet. The animals we eat must consume large amounts of plant matter to grow their bodies. It takes many pounds of plants to build a pound of herbivore muscle. It takes between three and twelve pounds of feed protein to produce a pound of meat, egg, or milk protein.[8] Protein is protein—the molecules are the same and one is not superior to the other. Consider our largest primate cousin, the mountain gorilla, whose males grow to 450 pounds of brawn on a vegan diet.

What are we doing about the meat problem? Not much. If human overpopulation is "the elephant in room" of policy reform, human meat consumption is the dinosaur under the desk of environmental reform. Time and again, one finds dietary themes missing in action among lists of "what you can do" to take personal responsibility for combating climate change. In May 2008, a global warming exhibit at Washington, D.C.'s Botanical Gardens featured over a dozen six-foot-diameter globes with various environmental themes: transportation, recycling, hydropower, the Kyoto Protocol, composting, walking, wind power, junk waste, cool roofs, shopping with reusable bags, bringing your own bag, planting a garden, etc. There was no mention of meat.

In the film *An Inconvenient Truth,* the need to curb meat consumption was too inconvenient a truth for Al Gore to mention. To its credit, however, *The Live Earth Global Warming Survival Handbook,* the companion book to the Live Earth concerts that the former vice president helped organize, lists "refusing meat" as "the single most effective thing you can do to reduce your carbon footprint."[9]

At a science education conference, I attended a plenary lecture by a climate change expert from the University of Montana who was one of the 450 or so lead authors on the Intergovernmental Panel on Climate Change, which was awarded the 2007 Nobel Peace Prize. The speaker didn't sugarcoat the crisis, and he mentioned several steps to address rising greenhouse gas emissions, such as biking or walking to work, supporting wind power, and addressing human overpopulation. But, dismayingly, he overlooked the enormous contribution of animal agriculture to global climate change. Speaking to about seven hundred biology teachers, he missed a huge opportunity to edify a large body of influential citizens.

Should legislation to curb meat consumption be enacted? It wouldn't be unprecedented. Governments during World War II didn't ask people to eat less—they enforced it with rations. Global warming may seem less urgent than WWII, but the consequences could be more grave. According to some climate change models, an average global temperature increase of 4°C (7.2°F)—a rather conservative prediction based on current trends—would result in huge swaths of the planet being uninhabitable for humans due to expanded desert habitats, rising sea levels, and the inability to grow food. The potential consequences include mass migration, and cities abandoned, and a projected human population reduced to one billion by 2099.[10] It is more likely to come true if we continue to ignore two of the root causes of climate change: human population growth and meat consumption. We can achieve only so much for the animals and the earth when we pontificate about being kind to them between bites of a cheeseburger.

We are beginning to notice the devastating impact of meat production. On the heels of a ten-year study of over half a million middle-aged and elderly Americans that linked higher consumption of red meat to increased death rates from heart disease and cancer, journalists acknowledged that tamping down greenhouse gas emissions was one of the benefits of weaning ourselves off meat.[11] The United Nations' climate chief, Rajendra Pachauri, who chairs the UN's Intergovernmental Panel on Climate Change, has publicly urged people to consider eating less meat as a way of combating global warming.[12]

For those unmoved by ecological concerns, there are some good personal reasons we might all follow Pachauri's advice. An estimated 76 million Americans are stricken with a food-borne illness every year; worldwide, some 2 million children die annually from contaminated food.[13] Animal products—or more specifically the pathogens that grow on contaminated animal tissues—top the list of causes.[14]

Sometimes, plants take the blame. When a batch of tomatoes, spinach, or peanuts causes an outbreak of food-borne illness, as they have recently in the United States, the public gets the mistaken impression that these plants are the direct culprits. But they are only the vectors or transmitters of pathogens that came from animals. As Dr. Neal Barnard, president and founder of the Physicians Committee for Responsible Medicine explains: "Salmonella are intestinal bacteria, and tomatoes have no intestine... these germs come from chicken and cow feces that contaminate waterways used for irrigation and contaminate kitchen counters and grocery store shelves."[15] To date, few media stories mention animal agriculture as the root cause of these outbreaks.

Nobody is suggesting that avian or swine flu are plant-borne illnesses, but what most consumers don't know is that there would be no swine flu epidemic without pig farms.[16] The same goes for avian flu and intensive chicken farming. Dr. Michael Greger, author of *Bird Flu: A Virus of Our Own Hatching* (2006) and director of public health and animal agriculture for The Human Society of the United States (HSUS), sees factory farms as "the perfect storm" environment for the emergence and spread of zoonoses. A zoonosis is any disease that spreads from one kind of animal to another—or from animals to humans or vice versa; zoonoses are usually spread by farm workers and the transport of livestock.[17]

To date, we have not been devastated by any of the zoonoses that continue to surface; we have not recently seen anything like the great influenza pandemic of 1918, which took between 20 and 40 million human lives. But today's human population and our global mobility mean that, in the words of Dr. Michael Osterholm, director of the U.S. Center for Infectious Disease Research and Policy, "an influenza pandemic of even moderate impact will result in the biggest single human disaster ever—far greater than AIDS, 9/11, all wars in the 20th century and the recent tsunami combined."[18]

Currently, global meat consumption is increasing. In 1983 the world consumed 152 million tons of meat a year. By 1997 it rose to 233 million tons. The FAO estimates that by 2020, world consumption could top 386 million tons of pork, chicken, beef, and farmed fish. Those are not facts but predictions, assuming current trends continue. They need not. We can choose to eat a lot fewer animals, or none at all.

New Days, New Ways

Human overpopulation and meat consumption are reversible. One of humankind's noblest traits is our capacity to change, and our moral consideration of animals is one of the more actively developing fields of thought. A century ago there weren't enough books on animal ethics to fill a desk drawer. Today there are some 1,500. There are also at least twelve academic journals in the field of animal ethics, and currently over two hundred courses taught at American and Canadian universities.[19] Animal law is a thriving discipline, with seventy-eight courses currently offered at American universities and colleges, according to the Animals & Society Institute. The accelerating pace of new legislation for animals may be attributable in part to a growing cadre of graduating attorneys who consider animals worthy of legal protection. The HSUS reports a record ninety-one animal protection laws passed in the United States in 2008, eclipsing the previous record of eighty-six new laws in 2007. Before 1985, there were no laws supporting an American student's right not to harm animals for a classroom exercise; today, fifteen states have enacted laws or policies that support students' conscientious objection to dissection.

The proliferation and expansion of animal protection organizations also reflects society's growing moral concern for animals. The online World Animal Net lists over 16,000 organizations in 173 countries. In 1970, The HSUS had 30,000 members and an annual budget of about $500,000; today, the organization is supported by over 10 million members and operates on a budget exceeding $120 million. People for the Ethical Treatment of Animals (PETA) has become virtually a household name in much of the Western world. Founded in 1980, its membership today totals over two million.

Since its formation in 1993, the European Union (EU), currently with twenty-seven member countries, has been leading the way on progressive animal legislation. In 1997 the EU officially recognized animals as sentient beings able to feel pain and emotion.

European Union members are required to comply with all EU legislation, but there is nothing to prevent them from independently enacting more far-reaching laws. In April 2008 the Swiss federal parliament passed sweeping legislation upholding some of animals' most basic rights, including the importance for social animals to interact with their kind. Its provisions prohibit farmers from tethering horses, sheep, or goats, or keeping cows and pigs, in areas with hard floors; and require dog *guardians* (the kinder way to refer to "owners") to pass a test of competence to care for animals. It also seems that the Swiss legislators were aware of recent studies on fish cognition and awareness, for the new law stipulates that social species such as goldfish not be kept alone, and requires anglers to complete a course on catching fish in a way that's more humane.[20] Norwegian law has included fishes in its Animal Welfare Act since 1974, and is close to granting fish protection on par with other vertebrates.[21] That fishes are beginning to gain the attention of welfare legislators illustrates the influence of scientific knowledge, and common sense, on policy.

Smaller jurisdictions are also passing compassionate legislation. As of October 2005, a municipal law was passed in Rome that requires guardians to regularly exercise their dogs, bans tail-docking for esthetic reasons, bans keeping goldfish in spherical bowls, and protects fish or any other animal from being given away as fairground prizes. Earlier, a nationwide law was passed in Italy making the abandonment of domestic animals a crime, punishable by a jail sentence. (Animal rights groups estimate some 150,000 to 200,000 dogs and cats are abandoned in Italy each year.) In the wake of the abandonment prohibition, several city councils were inspired to add their own animal welfare rules; Turin, for example, now requires a minimum of three daily walks for dogs. What I especially like about these changes is that they recognize animals' capacity not just to avoid suffering, but also to enjoy life.

In 1971, 55 percent of Spaniards claimed to be interested in bullfighting. By 2006, only 27 percent did.[22] As of May 2009, an anti-bullfighting platform named Prou had collected 180,000 signatures in a bid to outlaw bullfighting in Catalonia, a region of Spain with seven million residents. That's nearly four times the required minimum to stage a debate on the issue in the region's parliament. The decline in interest has shut down most bullrings in Catalonia except for one in Barcelona, maintained for tourists.[23] Currently, bullfights and bull runs are the only exemptions to a Catalan animal protection law that prohibits public spectacles that cause animal suffering. In 2007, a European Court of Human Rights vote in Strasbourg that would have declared bullfighting a form of torture narrowly fell short. This will likely not be the last time Strasbourg votes on the matter.

In the United States, the more popular blood sport of hunting is declining. In its 2006 survey, the Fish and Wildlife Service estimates that 12.5 million Americans hunted in 2006. That is a decline of about 4 percent, or half a million hunters, since the previous FWS survey five years earlier, and a 10 percent decline from the survey in 1996. Suspected causes for the wane include loss of land to urbanization and the perception that hunting is too time-consuming and costly. Neither of these would-be causes accounts for a 13 percent increase in wildlife *watching* since 1996.[24]

Even the hallowed halls of biomedical research are coming under growing legislative pressure. In the German town of Bremen the local parliament has called for the termination of experiments on live animals, in particular those of a neuroscientist whose work included placing electrodes in monkeys' brains.[25] In 2008, a lower Swiss court banned two invasive monkey experiments, ruling that the experiments would violate Swiss law requiring that a benefit to society must be weighed against the burden to animals before any animal procedures be approved, and also that the "dignity of creatures" be considered.[26] While the dignity issue was not raised in support of the court's decision, the court agreed that society was not likely to reap benefits from the research, which had been granted three years' funding by the Swiss National Science Foundation.

The most numerically widespread abuses of animals by far occur in their procurement as food for humans. Because of the scale and inertia of animal agriculture industries, they are among the most resistant to legislative reform, but they, too, are starting to shift.

The 2008 elections in California ushered in Proposition 2, which bans three of the worst factory farming practices: battery cages for chickens, veal crates, and gestation crates for pigs. All three must be abolished statewide by 2015, and the phase-out has begun.

Veal crates, gestation crates, and battery cages (industrial-scale, stacked and rowed caging systems containing several hens per bread-box-sized cage for collecting their eggs) are already banned in Europe. Norway's Parliament banned castration of piglets in 2009, the same year that major Dutch supermarkets pledged to stop selling meat from piglets castrated without anesthesia.[27] Since 1987, Great Britain has banned the so-called forced molting of hens. This relatively benign-sounding term refers to the practice of starving caged chickens for up to fourteen days, which causes the hens to lose all their feathers and to stop ovulation. This allows a farmer to streamline egg production and to save on feed costs. Starving hens remains legal in the United States, where, at any given time, according to the group United Poultry Concerns, some six million hens are being systematically starved.[28]

With the passage of Proposition 2, California became the fifth state to outlaw gestation crates—joining Florida, Arizona, Oregon, and Colorado—and the third to get rid of veal crates (after Arizona and Colorado). Perhaps most significantly, it became the first state to ban battery cages for laying hens, who are killed in far greater numbers than either pigs or calves. All told, Proposition 2 will spare over twenty million animals per year from some of the worst abuses of factory farming.

The significance of this legislation is that it acknowledges the intrinsic value of an animal's life. It says that a majority of voting Californians are willing to pay a little more to grant an animal a better life. It says that farmed animal well-being is no longer the concern of "bunny-huggers," but rather of citizens (and consumers) of conscience. Proposition 2 is a harbinger of widespread political change from the country's most populous state, and the nation's legislative bellwether. As this book goes to press, California has become the first state to bar tail-docking of dairy cows.

On May 20, 2008, the United States Department of Agriculture announced a complete ban on the slaughter for human consumption of "downer cattle"—animals who are too sick or severely injured to stand and walk unassisted. This move followed directly from an undercover investigation conducted by The HSUS that documented workers at a slaughter plant in Chino, California, pushing and jabbing downed cattle with forklifts, using chains to drag them across ridged concrete floors, and other abuses. Later conviction of the pen manager marked a legal milestone because animals raised for meat, eggs, and milk in the United States are normally denied—owing to legal loopholes or by virtue of being kept out of public sight—the most basic protections afforded other animals. In a public statement, HSUS president Wayne Pacelle urged Congress, industry, and consumers to confront the possibility that such problems are systemic and occur at other slaughter plants around the country.

There may be no more significant an indicator of emerging concern for animals than the formation of a political party dedicated to the animals' cause. The primary goal of the Dutch Party for the Animals (PftA) is simple: to create a society that treats animals with respect. Just three years after its formation in 2002, the PftA won two seats in the 150-seat Dutch House of Representatives. In 2007, PftA won nine seats in eight of the nation's twelve provinces, as well as one of the seventy-five seats in the Dutch Senate. As of June, 2009, the PftA has gained four percent of the popular vote and is on the cusp of winning one of the nation's twenty-five seats in the European Parliament.[29]

A rescued cow at a sanctuary enjoys a neck rub. (Photo courtesy of Connie Pugh.)

The changes we are now seeing on both sides of the Atlantic signal a growing recognition that animals matter, not just as species but as individuals. These are significant advances for a species that for most of its history has taken what it could without pause for the consequences for the vanquished. Or, ultimately, for the victors.

Second Nature

These changes signal a paradigm shift in our relationship to animals. As we continue to discover that animals have rich perceptual, emotional and cognitive lives, and that their lives are as valuable to them as ours are to us, we also realize that we cannot continue to treat them brutally. The era of our First Nature—in which we view animals as things to be used and taken for shortsighted gains—is coming to an end. Its downfall is inevitable because animal exploitation is unsustainable on our finite planet with a growing human population. The new era is grounded in science and driven by ethics. It is an emergent, less selfish worldview that grants animals the respect and consideration they're due. I call it Second Nature.

Embracing our Second Nature is an act of cultural change. We can take heart from the fact that moral and cultural change occur much, much faster than evolutionary change. Just two centuries ago the slave trade was still in full swing. In the past century, the suffrage movement led to women's voting rights, the Civil Rights movement gave equal rights (or brought them much closer) to black Americans, and Apartheid was dismantled in South Africa. Since World War II ended in 1945, centuries of British, French, Dutch, Spanish, and Portuguese colonialism have all but ended, Japan has thrived without a military force, Russia has abandoned its long experiment with communism, and legalized racism, homophobia, and subordination of women have been greatly reduced in America.[30]

The next great social advance for humankind is the establishment of basic freedoms for sentient animals.

How will this advance come about? It has already begun; witness the legislative changes summarized in the previous section. I also believe it will occur through the actions of a growing number of individuals who realize that exploiting animals is not good for their personal health or for the environment we all depend on. Each of us has the choice to make the world better. Such choice is immediate, and empowering. It is what inspired Anne Frank to write the words in this chapter's epigraph, "How wonderful it is that nobody need wait a single moment before starting to improve the world."

To choose to act mindfully and compassionately toward our animal counterparts is to be aware of the consequences of what we buy and what we eat. Whenever we purchase something, we are effectively telling the manufacturer of the product to *do it again.* This simple principle of supply and demand applies to all purchases, such as food, clothing, technology, pharmaceuticals, and personal care products.

This is at once sobering and empowering. Sobering because even though most consumers would cringe at what goes on in many animal industries—conveniently hidden from view in factory farms, slaughterhouses, fur farms, and testing labs—we endorse them with every dollar we spend on their products. Empowering, because we can immediately remove our support by the simple act of choosing to spend our money elsewhere. The opportunity to vote in political elections comes around only every few years. Voting with one's wallet happens on a daily basis.

Refusing to buy the products of animal cruelty is not just a passive act of omission. For consumers in developed nations—which use, per capita, more of these products than anywhere else—there are alternatives to virtually all products derived from animals. By buying something else, we put in a vote for companies that provide these more humane, sustainable alternatives. And it's easy. I switched from cow's milk to soy milk twenty years ago. I didn't do it because I had lost the taste for cow's milk. I did it because I did not agree with the treatment of cows and calves in the dairy industry. Taste is the most malleable of the senses, and within days mine had adjusted to the delicious sensation of soy milk—which today is available in practically every supermarket. The same goes for meat. I didn't give up meat because I didn't like it. My kitchen now contains hot dogs, sausages, meatballs, cold cuts, and four kinds of burgers, none of which contain any animal products.

Going vegan is the holy grail of personal activism for animals. When we stop buying and consuming animal products, we remove ourselves from any complicity in the institutionalized harms done to animals to produce those products. One needn't become a veg*n (vegetarian or vegan) overnight to make a positive impact. Any decrease in the amount of meat eaten is to the good. Meat production costs money, so if demand drops, economics dictate that the supply must follow suit. However, every time an ingrained habit is reinforced it becomes harder to break, so it is easier to give up meat "cold turkey," if you will, than in stages. You will also more quickly realize collateral benefits, such as weight loss (if you are heavy), lowered risk of heart disease, cancer or diabetes, and lowered risk of exposure to food-borne pathogens. And you can feel good in the knowledge that you are helping animals, and the planet.

There is another, subtler reason to shun cruel animal products for compassionate alternatives: violence against nonhumans goes "hand in paw" with violence against humans.

There is a growing stack of published studies linking cruelty to animals with societal violence. In a series of surveys published in the 1970s and 1980s by researchers at Yale University, strong associations were found between adult criminal behavior and childhood histories of animal cruelty. Such histories showed up in one-quarter of aggressive men who were in prison compared to none of a random sample of nonincarcerated men.[31] At a maximum security facility in Florida, the proportion of forty-five violent offenders who had committed past acts of cruelty to animals was significantly higher (56 percent) than that of forty-five nonviolent offenders (20 percent).[32] When cultural anthropologist David Levinson surveyed violence against women in ninety different human societies around the world, he found that victims were significantly more likely to be permanently injured, scarred, or killed by their husbands in societies in which animals were treated cruelly. Finally, an extensive analysis of 581 American counties with and without slaughterhouses (taking into account such variables as poverty levels and the number of males) found that, compared to other industries, slaughterhouse employment increases arrest rates for violent crimes, rape, and other sex offenses, presumably because the worker is desensitized to violence and cruelty.[33] These studies are just a small sampling of a considerable database linking animal abuse to interpersonal violence. Whereas animal cruelty was once treated as a low-priority problem, many police forces now take it seriously, because it signals a danger to society.

There is more to lose than gain in denying animals their rightful place on earth. When we treat other feeling creatures cruelly, we are more prone to treat other people that way. As Henry Salt explained in his 1892 book *Animals' Rights: Considered in Relation to Social Progress,* the movement to emancipate rights is a movement for human improvement.[34] If we heard of an alien civilization that takes babies away from mothers, eats the babies, and also consumes the mother's milk, we would probably not want to meet them. We stand to live in better, more caring societies when we treat all feeling individuals with compassion and respect.

Conclusion

The pianist Glenn Gould related an unhappy childhood memory of going fishing with his neighbor's family. "We went out in the boat and I was the first to catch a fish. And when the little perch came up and started wiggling about, I suddenly saw the scene entirely from the fish's point of view. I said: 'I want to put my fish back in the water'...The father pushed me back into my seat, probably for the sensible reason that I was rocking the boat [a phrase swollen with double meaning]. I've been violently anti-fishing ever since. To be really consistent, I should be a vegetarian."[35] Upon his death at fifty, Gould left half of his estate to the Society for the Prevention of Cruelty to Animals (SPCA).

Gould was expressing one of the most noble of emotions: empathy. He knew instinctively that fishes suffer when they are gaffed and removed from the water to suffocate. And it mattered to him because he knew it mattered to the fish.

As we've seen, fishes and other vertebrate animals have inner lives. As individuals with sensations, perceptions, emotions, and awareness, they experience life. Having the capacity to remember past events, and to anticipate future ones, animals' lives are not merely a series of now-moments; by showing that animals have ambient emotional states, we show that their lives play out like a moving tapestry, and they can go better or worse according to their circumstances. As active participants in dynamic communities teeming with other life forms, animals benefit by being on the ball, and learning from their experiences. Many live in rich social networks, where individuals benefit by forming friendships and by cooperating with others.

These capacities endow animals with interests of their own. They are not just living things; they are *beings with lives.* And that makes all the difference in the world. The next time you are outside, I encourage you to notice the first bird you see. Make a mental note that you are beholding a unique individual with personality traits, an emotional profile, and a library of knowledge built on experience. You may be looking at a majestic bald eagle or an ordinary house sparrow. It makes no difference—what you are witnessing is not just biology, but a biography.

When we make compassionate personal choices according to how they affect another, we are practicing Second Nature. The distinguished American biologist Edward O. Wilson coined the term *biophilia* to describe the connections that human beings unconsciously seek with the rest of life. Biophilia is a beautiful concept; it is the natural affinity—literally the love for life—that we may feel for a forest, the sound of ocean waves rolling onto a beach, the open sky, or a butterfly. Second Nature is a conscientious form of biophilia. It happens when we extend our affinity for nature to individuals—recognizing that they have lives of value and that they want to live as much as we do. Second Nature sees animals living rich sensory lives with hardships and rewards—like our lives. Second Nature understands the goodness built into animal societies, which, like ours, function smoothest when individuals get along and cooperate. Second Nature celebrates wild existence as a sensory odyssey not as the tedium of a constant struggle. Second Nature draws the connection between a slab of meat at the supermarket and a cow, a pig, or a chicken who not long ago lived and breathed. Second Nature is, in Harriet Beecher Stowe's immortal words, "the hand of benevolence . . . everywhere stretched out, searching into abuses, righting wrongs, alleviating distresses, and bringing to the knowledge and sympathies of the world the lowly, the oppressed, and the forgotten."[36] Extending our empathy and concern toward all who experience the ups and downs of life is neither strange nor radical. It is, after all, Second Nature.

notes

Part I

1. Ghirlanda, S., et al. "Chickens prefer beautiful humans." *Human Nature* 13 (2004): 383-389.

Chapter One

1. Helvétius, C.-A. *On the Mind* (*De l'esprit*, 1758), p. 276, quoted in Rousseau, J.-J., *Autobiographical, Scientific, Religious, Moral, and Literary Writings* (2006), vol. 12 of *The Collected Writings of Rousseau*. Hanover, NH: Dartmouth College Press, p. 211.
2. McCracken, G. F. "Locational memory and mother-pup reunions in Mexican free-tailed bat maternity colonies." *Animal Behaviour* 45 (1993): 811-813.
3. Balcombe, J. P. "Vocal recognition of pups by mother Mexican free-tailed bats, *Tadarida brasiliensis mexicana*." *Animal Behaviour* 39 (1990): 960-966.
4. Food and Agriculture Organization of the United Nations (FAO). "Livestock's long shadow: Environmental issues and options." Rome: FAO, 2006.
5. Carbone, L. *What Animals Want: Expertise and Advocacy in Laboratory Animal Welfare Policy*. Oxford and New York: Oxford University Press, 2004.
6. Sadler, K. C. "Mourning dove harvest." *Ecology and Management of the Mourning Dove*, edited by T. S. Baskett, et al. A Wildlife Management Institute Book. Harrisburg, PA: Stackpole Books, 1993.

Chapter Two

1. Tolle, E. *The Power of Now: A Guide to Spiritual Enlightenment.* Novato, CA: New World Library, 1999.
2. Primatt, H. *A Dissertation on the Duty of Mercy and the Sin of Cruelty to Brute Animals*... London: T. Cadell, J. Dodsley, etc., 1776.
3. Webster, J. "Animal sentience and animal welfare: What is it to them and what is it to us?" *Applied Animal Behaviour Science* 100 (2006):1-3.
4. Loeffler, K. "Schmerzen und Leiden beim Tier" [Pain and suffering in animals]. *Berliner und Münchener Tierärztliche Wochenschrift* (August 1, 1990) 103 (8): 257-61.
5. Silverman, J. "Sentience and sensation." *Lab Animal* 37 (2008): 465-467.
6. Broom, D. "Cognitive ability and sentience: Which aquatic animals should be protected?" *Diseases of Aquatic Organisms* 75 (2006): 99-108.
7. Bekoff, M. "Cognitive ethology and the treatment of nonhuman animals: How matters of mind inform matters of welfare." *Animal Welfare* 3 (1994): 75-96.
8. Rollin, B. *The Unheeded Cry: Animal Consciousness, Animal Pain, and Science.* New York: Oxford University Press, 1989, p. 238.
9. Sachar, E. S., and Sachar, E. J. "Hormonal changes in stress and mental illness." *Neuroendocrinology*, Sunderland, MA: Sinauer Associates, 1980, p. 477.
10. Balcombe, J. P., et al. "Laboratory routines cause animal stress." *Contemporary Topics in Laboratory Animal Science* 43(6) (2004): 42-51.
11. Sharp, J. L., et al. "Does witnessing experimental procedures produce stress in male rats?" *Contemporary Topics in Laboratory Animal Science* 41 (2003): 8-12.
12. Farm Sanctuary. *Sentient Beings: A Summary of the Scientific Evidence Establishing Sentience in Farmed Animals.* Watkins Glen, NY: Farm Sanctuary, 2007.
13. Jones, G. "Echolocation." *Current Biology* 15 (2005): R484-R488.
14. Barclay, R. M. R. "Interindividual use of echolocation calls: Eavesdropping by bats." *Behavioral Ecology and Sociobiology* 10 (1982): 271-275.
15. Balcombe, J. P., and M. B. Fenton. "Eavesdropping by bats: The influence of echolocation call design and foraging strategy." *Ethology* 79 (1988): 158-166.
16. Heyers, D., et al. "A visual pathway links brain structures active during magnetic compass orientation in migratory birds." *PLoS ONE* 2(9) (2007): e937. doi:10.1371/journal.pone.0000937.

17. Begall, S., et al. "Magnetic alignment in grazing and resting cattle and deer." *Proceedings of the National Academy of Sciences* 105(36) (2008): 13451-13455.
18. Rooney, N., and J. W. S. Bradshaw. "Social cognition in the domestic dog: Behaviour of spectators towards participants in interspecific games." *Animal Behaviour* 72 (2006): 343-352.
19. Uhlenbroek, C. *Talking with Animals*. London: Hodder & Stoughton, 2002.
20. Downer, J. *Weird Nature*. London: British Broadcasting Corporation, 2002.
21. Simmons, J. A. "Formation of perceptual objects from the timing of neural responses: Target-range images in bat sonar." *The Mind-Brain Continuum: Sensory Processes*, edited by R. R. Llinás and P. S. Churchland, Cambridge, MA: MIT Press, 1996, pp. 219-50.
22. Mann N. I., Dingess, K. A., Slater, P. J. B. 2005. "Antiphonal four-part synchronized chorusing in a Neotropical wren." *Biology Letters* 2: 1-4.
23. Slabbekoorn, H., and M. Peet. "Ecology: Birds sing at a higher pitch in urban noise." *Nature* (2003): 424:267.
24. Moss, C. *Portraits in the Wild: Animal Behavior in East Africa*. 2nd ed. Chicago: University of Chicago Press, 1982.
25. Fountain, H. "Squirrels that predict the future." *New York Times*, January 2, 2007.
26. Platt, M. "Animal cognition: Monkey meteorology." *Current Biology* 16 (2006): R464-R466.
27. Royte, E. *The Tapir's Morning Bath: Mysteries of the Tropical Rain Forest and the Scientists Who Are Trying to Solve Them.* Boston: Houghton Mifflin, 2001.
28. Read, M. A., et al. "Satellite tracking reveals long distance coastal travel and homing by translocated estuarine crocodiles, *Crocodylus porosus*." PLoS ONE 2(9) (2007): e949. doi:10.1371/journal.pone.0000949
29. Pozis-Francois, O., et al. "Social play in Arabian babblers." *Behaviour* 141 (2004): 425-450.
30. Bekoff, M. "Social play and play soliciting by infant canids." *American Zoologist* 14 (1974): 323-340.
31. Martin, C. J., et al. "Feedback control in active sensing: Rat exploratory whisking is modulated by environmental contact." *Proceedings of the Royal Society B: Biological Sciences* 274 (2007): 1035-1041. Ritt, J. T., et al. "Embodied information processing: Vibrissa mechanics and texture features shape micromotions in actively sensing rats." *Neuron* 57 (2008): 599-613.

32. Wong, B. B. M. "Superior fighters make mediocre fathers in the Pacific blue-eye fish." *Animal Behaviour* 67 (2004): 583-590.
33. Angier, N. "Who Is the Walrus?" *New York Times,* May 20, 2008.
34. Murphy, W. J., et al. "Molecular phylogenetics and the origins of placental mammals." *Nature* 409 (2001): 614-618.
35. Burn, C. C. "What is it like to be a rat? Rat sensory perception and its implications for experimental design and rat welfare." *Applied Animal Behaviour Science* 112 (2008): 1-32.
36. Hurst, J. L., et al. "Individual recognition in mice mediated by major urinary proteins." *Nature* 414(6864) (2001): 631-634.
37. Manning, A., and M. S. Dawkins. *An Introduction to Animal Behaviour.* 5th ed. Cambridge, U.K.: Cambridge University Press, 1998.
38. Ehman, K., and M. Scott. "Urinary odor preferences of MHC congenic female mice, *Mus domesticus*: Implications for kin recognition and detection of parasitized males." *Animal Behaviour* 62 (2001): 781-789.
39. Kavanau, J. L. "Behavior of captive white-footed mice." *Science* 155 (1967): 1623-1639.
40. Gladwell, M. *Blink: The Power of Thinking without Thinking.* Boston: Little, Brown, 2005.
41. Gallagher, S. "How to undress the affective mind: An interview with Jaak Panksepp." *Journal of Consciousness Studies* 15 (2008): 89-119.
42. Gould, J. L., and C. G. Gould. *The Animal Mind.* New York: Scientific American Library, 1994.
43. Richard, P. B. *Les Castors* [Beavers]. Poitiers, France: Balland, 1980.
44. Griffin, D. R. *Animal Minds: Beyond Cognition to Consciousness.* 1992. Chicago: University of Chicago Press, 2001, p. 106.
45. Humphrey, N. K. "Nature's Psychologists." *Consciousness and the Physical World: Edited Proceedings of an Interdisciplinary Symposium on Consciousness Held at the University of Cambridge in January 1978.* Edited by B. D. Josephson and V. S. Ramachandran. New York: Pergamon, 1980, p. 80.

Chapter Three

1. Marzluff, J. M., and T. Angell. *In the Company of Crows and Ravens.* New Haven, CT: Yale University Press, 2005, p. 80.
2. Barkley, C. L., and L. Jacobs. "Sex and species differences in spatial memory in food-storing kangaroo rats." *Animal Behaviour* 73 (2007): 321-329.
3. Manier, J. "Empathy for one's fellow chimp: Experts now think the apes may relate to each other in very human ways." *Chicago Tribune,* March 23, 2007.

4. McRae, F. "I'm the chimpion! Ape trounces the best of the human world in memory competition." *Daily Mail* (U.K.), January 26, 2008.
5. Spinney, L. "When chimps outsmart humans." *New Scientist* 190 (2006): 48-49.
6. Hooper, R. "Help me, please." *New Scientist* 193 (2007): 15 .
7. De Rohan, A. "Deep thinkers: The more we study dolphins, the brighter they turn out to be." *Guardian* (U.K.), July 3, 2003.
8. Linden, E. *The Octopus and the Orangutan: New Tales of Animal Intrigue, Intelligence, and Ingenuity*. New York: Dutton, 2002.
9. Mendes, N., et al. "Raising the level: Orangutans use water as a tool." *Biology Letters* 3 (2007): 453-455.
10. Clayton, N. S., and A. Dickinson. "Episodic-like memory during cache recovery by scrub jays." *Nature* 395 (1998): 272-274.
11. Henderson, J., et al. "Timing in free-living rufous hummingbirds, *Selasphorus rufus*." *Current Biology* 16 (2006): 512-515.
12. Dere, E., et al. "Integrated memory for objects, places, and temporal order: Evidence for episodic-like memory in mice." *Neurobiology of Learning and Memory* 84 (2005): 214-221.
13. Balcombe, J. P. "Laboratory environments and rodents' behavioural needs: A review." *Laboratory Animals* 40 (2006): 217–235.
14. Neuringer, A. J. "Animals respond for food in the presence of free food." *Science* 166 (1969): 399-401.
15. Ferkin, M. H., et al. "Meadow voles, *Microtus pennsylvanicus*, have the capacity to recall the 'what,' 'where," and 'when' of a single past event." Animal Cognition 11 (2008): 147-159.
16. Brown, C., et al., eds. *Fish Cognition and Behavior*. Oxford: Wiley-Blackwell, 2006.
17. Milinski, M., et al. "Do sticklebacks cooperate repeatedly in reciprocal pairs?" *Behavioural Ecology and Sociobiology* 27 (1990): 17-21. Milinski, M., et al. "Tit-for-tat: Sticklebacks (*Gasterosteus aculeatus*) 'trusting' a cooperating partner." *Behavioral Ecology* 1 (1990): 7-12.
18. Ward, A. J. W., and P. J. B. Hart. "Foraging benefits of shoaling with familiars may be exploited by outsiders." *Animal Behaviour* 69 (2005): 329-335.
19. Chase, A. R. "Music discriminations by carp (*Cyprinus carpio*)." *Animal Learning & Behavior* 29 (2001): 336-353.
20. Fisher, H. S., and G. G. Rosenthal. "Female swordtail fish use chemical cues to select well-fed mates." *Animal Behaviour* 72 (2006): 721-725.
21. Alfieri, M. S., and L. A. Dugatkin. "Cooperation and cognition in fishes." In *Fish Cognition and Behavior*, edited by C. Brown, et al. Oxford: Wiley-Blackwell, 2006.

22. Oliveira, R. F., et al. "Know thine enemy: Fighting fish gather information from observing conspecific interactions." *Proceedings of the Royal Society B: Biological Sciences* 265 (1998): 1045-1049.
23. Doutrelant, C., et al. "The effect of an audience on intrasexual communication in male Siamese fighting fish, *Betta splendens*." *Behavioral Ecology* 12 (2001): 283-286.
24. Lund, V., et al. "Expanding the moral circle: farmed fish as objects of moral concern." *Diseases of Aquatic Organisms* 75 (2007): 109-118.
25. Fernández-Juricic, E., et al. "Increasing the costs of conspecific scanning in socially foraging starlings affects vigilance and foraging behaviour." *Animal Behaviour* 69 (2005): 73-81.
26. Beran, M. J., et al. "A Stroop-like effect in color-naming of color-word lexigrams by a chimpanzee (*Pan troglodytes*)." *Journal of General Psychology* 134 (2007): 217-228.
27. Moss, C. *Portraits in the Wild: Animal Behavior in East Africa (Second Edition)*. Chicago: University of Chicago Press, 1982.
28. Gould, S. J. *The Mismeasure of Man*. New York: W. W. Norton, 1981.
29. Bentham, J. *An Introduction to the Principles of Morals and Legislation*. 1789. An authoritative edition edited by J. H. Burns and H. L. A. Hart. Oxford: Clarendon, 1996.

Chapter Four

1. Bekoff, M. *The Emotional Lives of Animals: A Leading Scientist Explores Animal Joy, Sorrow, and Empathy—and Why They Matter*. Novato, CA: New World Library, 2007.
2. Masson, J., McCarthy, S. *When Elephants Weep: The Emotional Lives of Animals*. New York: Delacorte Press. 1995.
3. Preuschoft, S., and J. A. R. A. M. van Hooff. "Homologizing primate facial displays: A critical review of methods." *Folia Primatologica* 65 (1995): 121–37.
4. Cheney, D. L., and R. M. Seyfarth. *Baboon Metaphysics: The Evolution of a Social Mind*. Chicago: University of Chicago Press, 2007.
5. Engh, A. L., et al. "Behavioral and hormonal responses to predation in female chacma baboons (*Papio hamadryas ursinus*)." *Proceedings of the Royal Society B: Biological Sciences* 273 (2006): 707-712.
6. Cabanac, M. "The experience of pleasure in animals." In *Mental Health and Well-being in Animals*, edited by F. D. McMillan, pp. 29-46. Ames, IA: Blackwell, 2005.
7. Renbourn, E. T. "Body temperature and the emotions." *Lancet* 2 (1960): 475-476.

8. Briese, E., and M. G. de Quijada. "Colonic temperature of rats during handling." *Acta Physiologica Latinoamericana* 20 (1970): 97-102.
9. Parr, L. A. "Cognitive and physiological markers of emotional awareness in chimpanzees (*Pan troglodytes*)." *Animal Cognition* 4 (2001): 223-229.
10. Wemelsfelder, F., et al. "The spontaneous qualitative assessment of behavioural expressions in pigs: First explorations of a novel methodology for integrative animal welfare measurement." *Applied Animal Behaviour Science* 67 (2001): 193-215.
11. Bekoff, M. *Minding Animals: Awareness, Emotions, and Heart.* Foreword by Jane Goodall. New York: Oxford University Press, 2002.
12. Nicastro, N., and M. J. Owren. "Classification of domestic cat (*Felis catus*) vocalizations by naïve and experienced human listeners." *Journal of Comparative Psychology* 117 (2003): 44-52.
13. De Waal, F. B. M. *Our Inner Ape: A Leading Primatologist Explains Why We Are Who We Are.* New York: Riverhead, 2005.
14. Fimrite, P. "Daring rescue of whale off Farallones." *San Francisco Chronicle*, December 14, 2005.
15. Bateson, M., and S. M. Matheson. "Performance on a categorisation task suggests that removal of environmental enrichment induces 'pessimism' in captive European starlings (*Sturnus vulgaris*)." *Animal Welfare* 16S (2007): 33-36.
16. Mason, G. J., and N. R. Latham. "Can't stop, won't stop: Is stereotypy a reliable animal welfare indicator?" *Animal Welfare* 13 (2004): S57–S69.
17. Asher, L, and M. Bateson, unpublished data.
18. Harding, E. J., et al. "Animal behaviour: Cognitive bias and affective state." *Nature* 427 (2004): 312–1312.
19. Bardo, M. T., et al. "Environmental enrichment decreases intravenous self-administration of amphetamine in female and male rats." *Psychopharmacology* 155(3) (2001): 278-84.
20. Sherwin, C. M., and I. A. S. Olsson. "Housing conditions affect self-administration of anxiolytic by laboratory mice." *Animal Welfare* 13 (2004): 33–9.
21. Pozis-Francois, O., et al. "Social play in Arabian babblers." *Behaviour* 141 (2004): 425-450.
22. McComb, K., et al. "African elephants show high levels of interest in the skulls and ivory of their own species." *Proceedings of the Royal Society B: Biology Letters* 2 (2006): 26-28.doi:10.1098/rsbl.2005.0400
23. Douglas-Hamilton, I., et al. "Behavioural reactions of elephants towards a dying and deceased matriarch." *Applied Animal Behaviour Science* 100 (2006): 87-102.
24. Pitman, T. "South Africa lifts ban on elephant culls." *The Guardian* (U.K.), May 2, 2008.

25. Ryan, M., and P. Thornycroft. "Jumbos mourn black rhino killed by poachers." *Sunday Independent* (U.K.), November 18, 2007.
26. Stachowski, K. "When bison grieve: notes from Montana's 'fair chase' hunt." *New West*, December 17, 2005.
27. Moss, C. *Portraits in the Wild: Animal Behavior in East Africa.* 2nd ed. Chicago: University of Chicago Press, 1982.
28. Marder, A. R., and J. M. Posage. "Treatment of emotional distress and disorders—pharmacologic methods." *Mental Health and Well-being in Animals*, edited by F. D. McMillan, pp. 159-166. Ames, IA: Blackwell, 2005.
29. Phillips, B. "The Nature of Elephants." *Memphis Flyer*, January 25, 2007.
30. Bradshaw, G. A., et al. "Elephant breakdown." *Nature* 433 (2005): 807.
31. Brüne, M., et al. "Psychiatric treatment for great apes?" *Science*, 306(5704) (2004): 2039. Bradshaw, G. A., et al. "Building an inner sanctuary: Complex PTSD in chimpanzees." *Journal of Trauma & Dissociation* 9 (2008): 9-34.
32. Hall, L., and A. J. Waters. "From property to person: The case of Evelyn Hart." *Seton Hall Constitutional Law Journal* 11(1) (2000).
33. Sandem, A. I., Braastad, B. O., and Bøe, K. E. "Eye white may indicate emotional state on a frustration-contentedness axis in dairy cows." *Applied Animal Behaviour Science* 79 (2002):1-10. Sandem, A. I., Braastad, B.O. "Effects of cow-calf separation on visible eye white and behaviour in dairy cows—A brief report." *Applied Animal Behaviour Science* 95 (2005):233-239.

Chapter Five

1. Bates, L. A., et al. "African elephants have expectations about the locations of out-of-sight family members." *Biology Letters* 4 (2008): 34-36.
2. Call, J., et al. "'Unwilling' versus 'unable': Chimpanzees' understanding of human intentional action." *Developmental Science* 7 (2004): 488-498.
3. Bräuer, J., et al. "All great ape species follow gaze to distant locations and around barriers. *Journal of Comparative Psychology* 119 (2005): 145-154.
4. Hare, B., et al. "Chimpanzees know what conspecifics do and do not see." *Animal Behaviour* 59 (2000): 771-785.
5. Premack, D., and A. J. Premack. *The Mind of an Ape.* New York: W. W. Norton, 1984.
6. Hostetter, A. B., et al. "Now you see me, now you don't: Evidence that chimpanzees understand the role of the eyes in attention." *Animal Cognition* 10 (2007): 55-62.

7. Whiten, A., and R. Byrne. "Tactical deception in primates." *Behavioral Brain Science* 11 (1988): 233–273.
8. Russell, J. L., et al. "Chimpanzee (*Pan troglodytes*) intentional communication is not contingent upon food." *Animal Cognition* 8 (2005): 263-272.
9. Gácsi, M., et al. "Are readers of our face readers of our minds? Dogs (*Canis familiaris*) show situation-dependent recognition of human's attention." *Animal Cognition* 7 (2004): 144-153.
10. Adachi, I., et al. "Dogs recall their owner's face upon hearing the owner's voice." *Animal Cognition* 10 (2007): 17-21.
11. Prior, H., et al. "Mirror-induced behavior in the magpie (*Pica pica*): Evidence of self-recognition." *PLoS Biology* 6(8) (2008), e202. doi:10.1371/journal.pbio.0060202.
12. Cheney, D. L., and R. M. Seyfarth. *Baboon Metaphysics: The Evolution of a Social Mind.* Chicago: University of Chicago Press, 2007.
13. Phillips, H. "Known unknowns." *New Scientist* 192 (2006): 28-31.
14. Foote, L. A., and J. D. Crystal. "Metacognition in the rat." *Current Biology* 17 (2007): 551-555.
15. Koch, J. "Behavioral science turns to dogs for answers." *Spiegel Online,* September 7, 2007.
16. LaVoie, A. "Primates expect others to act rationally." *Harvard University Gazette,* September 6, 2007.
17. Range, F., et al. "Selective imitation in dogs." *Current Biology* 17 (2007): 868-872.
18. Makowska, I. J., and D. L. Kramer. "Vigilance during food handling in grey squirrels, *Sciurus carolinensis.*" *Animal Behaviour* 74 (2007): 153-158.
19. Plath, M., et al. "Audience effect alters mating preferences in a livebearing fish, the Atlantic molly, *Poecilia Mexicana.*" *Animal Behaviour* 75 (2007): 21-29.
20. Dzieweczynski, T. L., et al. "Audience effect is context dependent in Siamese fighting fish, *Betta splendens.*" *Behavioral Ecology* 16 (2005): 1025-1030.
21. Rozell, N. "Her fat suit fools some wild ravens." *Alaska Science*, September 9, 2007.
22. Steinhart, P. *The Company of Wolves.* New York: Vintage, 1996.
23. Hediger, H. *Studies in the Psychology and Behaviour of Animals in Zoos and Circuses.* London: Butterworth's Scientific Publications, 1955.
24. De Waal, F. B. M. *Good Natured: The Origins of Right and Wrong in Humans and Other Animals.* Cambridge, MA: Harvard University Press, 1996.

25. De Waal, F. B. M. "Are we in anthropodenial?" *Discover* 18 (1997): 50-53.
26. Melis, A. P., et al. "Chimpanzees (*Pan troglodytes*) conceal visual and auditory information from others." *Journal of Comparative Psychology* 120 (2006): 154-162.
27. Hare, B., et al. "Chimpanzees deceive a human competitor by hiding." *Cognition* 101 (2006): 495-514.
28. Santos, L. R., et al. "Rhesus monkeys, *Macaca mulatta*, know what others can and cannot hear." *Animal Behaviour* 71 (2006): 1175-1181.
29. De Waal, F. B. M. *Our Inner Ape: A Leading Primatologist Explains Why We Are Who We Are*. New York: Riverhead, 2005.
30. Munn, C. A. "Birds that 'cry wolf'." *Nature* 391 (1986): 143-145.
31. Kemper, S. "Who's laughing now." *Smithsonian* 39 (2008): 76-84.
32. Cheney, D. L., and R. M. Seyfarth. *How Monkeys See the World: Inside the Mind of Another Species*. Chicago: University of Chicago Press, 1990.
33. Newby, J. "Cheating Chooks." Australian Broadcasting Corporation, March 10, 2002.
34. Bugnyar, T., and K. Kotrschal. "Leading a conspecific away from food in ravens (*Corvus corax*)?" *Animal Cognition* 7 (2004): 69-76.
35. Heinrich, B. *Mind of the Raven: Investigations and Adventures with Wolf-Birds*. New York: HarperCollins, 1999.
36. Bugnyar, T., and K. Kotrschal. "Observational learning and the raiding of food caches in ravens, *Corvus corax*: Is it 'tactical' deception?" *Animal Behavior* 64 (2002): 185-195.
37. Moss, C. *Portraits in the Wild: Animal Behavior in East Africa*. 2nd ed. Chicago: University of Chicago Press, 1982.
38. Steele, M. A., et al. "Cache protection strategies of a scatter-hoarding rodent: Do tree squirrels engage in behavioural deception?" *Animal Behaviour* 75 (2008): 705-714.
39. Gould, J. L., and C. G. Gould. *The Animal Mind*. New York: Scientific American Library, 1994.
40. Ibid.
41. Aeschbacher, A., and G. Pilleri. "Observations on the building behaviour of the Canadian beaver (*Castor Canadensis*) in captivity." In *Investigations on Beavers*, vol. 1, edited by G. Pilleri. Berne, Switzerland: Brain Anatomy Institute, 1983.
42. Gould and Gould, *The Animal Mind*.
43. Griffin, D. R. *Animal Minds: Beyond Cognition to Consciousness*. Revised and expanded. Chicago: University of Chicago Press, 2001, p. 106.

44. Blakeslee, S. "What a rodent can do with a rake in its paw." *New York Times,* March 26, 2008.
45. Katz, S. J. "Tool-using behavior of the pocket gopher, *Thomomys bottae* (*Geomyidae*)?" *Southwestern Naturalist* 25 (1980): 270-271.
46. Shuster, G., and P. W. Sherman. "Tool use by naked mole-rats." *Animal Cognition* 1 (1998): 71-74.
47. Grandin, T., and C. Johnson. *Animals in Translation*. Orland, Florida: Harcourt, 2005. p. 303.

Chapter Six

1. Palmer, M. E., et al. "Response of female cuttlefish *Sepia officinalis* (Cephalopoda) to mirrors and conspecifics: Evidence for signaling in female cuttlefish." *Animal Cognition* 9 (2006): 151-155.
2. Lindquist, E. D., and T. E. Hetherington. "Semaphoring in an earless frog: The origin of a novel visual signal." *Animal Cognition* 1 (1998): 1435-1448.
3. Grafe, T. U., and T. C. Wanger. "Multimodal signaling in male and female foot-flagging frogs *Staurois guttatus* (Ranidae): An alerting function of calling." *Ethology* 113 (2007): 772-781.
4. Holmes, B. "Close calls don't fool smart tree frogs." *New Scientist* 199 (2008): 18.
5. Dunlop, T. "Australian scientists decode whale sounds." Reuters, November 8, 2007.
6. Simmonds, M. P. "Into the brains of whales." *Applied Animal Behaviour Science* 100 (2006): 103-116.
7. Wilson, B. "Noisy herring." In *Encyclopedia of Animal Behavior*, 3 vols., edited by M. Bekoff, pp. 335-336. Westport, CT: Greenwood Press, 2004.
8. Wilson, B., et al. "Pacific and Atlantic herring produce burst pulse sounds." *Biology Letters* 10 (2003):1098.
9. Owen, J. "Herring break wind to communicate, study suggests." *National Geographic* online, November 10, 2003.
10. De Waal, F. B. M. "Storming the language barrier." *Nature* 452 (2008): 154.
11. Wynne, C. D. L. *Do Animals Think?* Princeton, NJ: Princeton University Press, 2004.
12. Gould, J. L., and C. G. Gould. *The Animal Mind.* New York: Scientific American Library, 1994.
13. Struhsaker, T. T. "Auditory communication among vervet monkeys (*Cercopithecus aethiops*)." In *Social Communication Among Primates,*

edited by S. A. Altmann, pp. 281-324. Chicago: University of Chicago Press, 1967.

14. Evans, C. S., and L. Evans. "Chicken food calls are functionally referential." *Animal Behaviour* 58 (1999): 307-319.
15. Palleroni, A., et al. "Do responses of galliform birds vary adaptively with predator size?" *Animal Cognition* 8 (2005): 200-210.
16. Templeton, C. N., et al. "Allometry of Alarm Calls: Black-capped Chickadees Encode Information About Predator Size." *Science* 308 (2005): 1934-1937.
17. Chester, C. *Providence of a Sparrow: Lessons from a Life Gone to the Birds.* New York: Anchor, 2002, p. 220.
18. Howard, L. *Birds as Individuals.* London: Collins, 1952.
19. Slobodchikoff, C. N., et al. "Semantic information distinguishing individual predators in the alarm calls of Gunnison's prairie dogs." *Animal Behaviour* 42 (1991): 713-719.
20. Millett, R. P., and J. P. Pratt. "Prairie dog language?" *Meridian Magazine,* May 19, 2005.
21. Frederiksen, J. K., and C. N. Slobodchikoff. "Referential specificity in the alarm calls of the black-tailed prairie dog." *Ethology Ecology & Evolution* 19 (1991): 87-99.
22. Slobodchikoff C. N., et al. "Prairie dog alarm calls encode labels about predator colors." *Animal Cognition 12* (2008): 435-439.
23. Lindsey, J. *Jane Goodall: 40 Years at Gombe: A Tribute to Four Decades of Wildlife Research, Education, and Conservation.* Produced in association with the Jane Goodall Institute. New York: Stewart, Tabori & Chang, 1999.
24. Gould and Gould, *The Animal Mind.*
25. Liebal, K., et al. "Use of gesture sequences in chimpanzees." *American Journal of Primatology* 64 (2004): 377-396.
26. Leavens, D. A., et al. "Referential communication by chimpanzees (*Pan troglodytes*)." *Journal of Comparative Psychology* 118 (2004): 48-57.
27. Menzel, C. R. "Unprompted recall and reporting of hidden objects by a chimpanzee (*Pan troglodytes*) after extended delays." *Journal of Comparative Psychology* 113 (1999): 426-434.
28. Page, G. *Inside the Animal Mind: A Groundbreaking Exploration of Animal Intelligence.* New York, Doubleday, 1999.
29. Pepperberg, I. M. "'Insightful' string-pulling in Grey parrots (*Psittacus erithacus*) is affected by vocal competence." *Animal Cognition* 7 (2004): 263-266.
30. Van Lawick, H. *Among Predators and Prey: A Photographer's Reflections on African Wildlife.* San Francisco: Sierra Club Books, 1986.

31. Parr, L. A., and F. B. M. de Waal. "Visual kin recognition in chimpanzees." *Nature* 399 (1999): 647-648.
32. Kojima, S., et al. "Identification of vocalizers by pant hoots, pant grunts and screams in a chimpanzee." *Primates* 44 (2003): 225-230.
33. Godard, R. "Long-term memory of individual neighbors in a migratory songbird." *Nature* 350 (1991): 228-229.
34. Howard, *Birds as Individuals.*
35. Moss, C. *Portraits in the Wild: Animal Behavior in East Africa.* 2nd ed. Chicago: University of Chicago Press, 1982.
36. Janik, V. M., et al. "Signature whistle shape conveys identity information to bottlenose dolphins." *Proceedings of the National Academy of Sciences (USA)* 103 (2006): 8293-8297.
37. Wanker, R., et al. "Vocal labeling of family members in spectacled parrotlets, *Forpus conspicillatus.*" *Animal Behaviour* 70 (2005): 111-118.
38. Schmid, R. E. "Nuthatches seem to understand chickadee." Associated Press, March 20, 2007.
39. Goodale, E., and S. W. Kotagama. "Vocal mimicry by a passerine bird attracts other species involved in mixed-species flocks." *Animal Behaviour* 72 (2006): 471-477.
40. Goodale, E., et al. "Birds of a different feather." *Natural History* 117 (2008): 25-28.
41. Fichtel, C. "Reciprocal recognition of sifaka (*Peopithecus verreauxi verreauxi*) and redfronted lemur (*Eulemur fulvus rufus*) alarm calls." *Animal Cognition* 7 (2004): 45-52.
42. Rainey, H. J., et al. "Hornbills can distinguish between primate alarm calls." *Proceedings of the Royal Society B: Biological Sciences* 271 (2004): 755-759.
43. Deecke, V. B., et al. "The vocal behaviour of mammal-eating killer whales: Communicating with costly calls." *Animal Behaviour* 69 (2005): 395-405.
44. BBC News. "Gecko 'begs' insect for honeydew." British Broadcasting Corporation, February 16, 2008.
45. Habersetzer, J. "Adaptive echolocation sounds in the bat *Rhinopoma hardwickei.*" *Journal of Comparative Physiology, A: Sensory, Neural, and Behavioral Physiology* 144 (1981): 559–566. Ratcliffe, J. M., et al. "Conspecifics influence call design in the Brazilian free-tailed bat, *Tadarida brasiliensis.*" *Canadian Journal of Zoology* 82 (2004): 966-971.
46. Gillam, E. H., and G. F. McCracken. "Variability in the echolocation of *Tadarida brasiliensis*: Effects of geography and local acoustic environment." *Animal Behaviour* 74 (2007): 277-286.

47. Tallarovic, S. K., and H. H. Zakon. "Electric organ discharge frequency jamming during social interactions in brown ghost knifefish, *Apteronotus leptorhynchus*." *Animal Behaviour* 70 (2005): 1355-1365.
48. Brumm, H., and P. J. B. Slater. "Animals can vary signal amplitude with receiver distance: Evidence from zebra finch song." *Animal Behaviour* 72 (2006): 699-705.
49. Fuller, R. A., et al. "Daytime noise predicts nocturnal singing in urban robins." *Biology Letters* 3 (2007): 368-370.
50. Yong, E. "Wing tones: Birds are going to extraordinary lengths to make themselves heard in cities." *New Scientist* 197 (2008): 33-35.

Chapter Seven

1. Margulis, L., and D. Sagan. "Marvellous microbes: From microbes to the human race, all organisms are equally evolved." *Resurgence* 206 (2001): 10–12.
2. Bäckhed, F., et al. "Host-bacterial mutualism in the human intestine." *Science* 307 (2005): 1915-1920.
3. Dawkins, R. *The Ancestor's Tale: A Pilgrimage to the Dawn of Evolution.* Boston: Houghton Mifflin, 2004.
4. Dugatkin, L. A. *Cooperation Among Animals: An Evolutionary Perspective.* Oxford: Oxford University Press, 1997.
5. Hamilton, W. D. "The genetical theory of social behaviour (I and II)." *Journal of Theoretical Biology* 7 (1964), 1–16, 17–32.
6. Trivers, R. L. "The evolution of reciprocal altruism." *Quarterly Review of Biology* 46 (1971): 35-57.
7. Locantore, J. "The naked truth about mole-rats." *Smithsonian Zoogoer*, May-June 2002.
8. De Kort, S. R., et al. "Food sharing in jackdaws, *Corvus monedula*: What, why and with whom?" *Animal Behaviour* 72 (2006): 297-304.
9. De Waal, F. B. M. *Our Inner Ape: A Leading Primatologist Explains Why We Are Who We Are.* New York: Riverhead, 2005.
10. Brosnan, S. F., et al. "Partner's behavior, not reward distribution, determines success in an unequal cooperative task in capuchin monkeys." *American Journal of Primatology* 68 (2006): 713-724.
11. Wilkinson, G. "Food sharing in vampire bats." *Scientific American* 262 (1990): 64-70.
12. Brown, C. R., et al. "Food-sharing Signals Among Socially Foraging Cliff Swallows." *Animal Behaviour* 42 (1991): 551–565.

13. Slobodchikoff, C. "Cooperation, not competition." *Reconnect with Nature Blog*, December 29, 2006. http://www.reconnectwithnatureblog.com/2006/12/index.html.
14. Corning, P. *Nature's Magic: Synergy in Evolution and the Fate of Humankind.* Cambridge, U.K.: Cambridge University Press, 2003.
15. Heinrich, B., and J. Marzluff. "Do common ravens yell because they want to attract others?" *Behavioral Ecology and Sociobiology* 28 (1991): 13-21.
16. Pays, O., et al. "Coordination, independence or synchronization of individual vigilance in the eastern grey kangaroo?" *Animal Behaviour* 73 (2007): 595-604.
17. Tebbich, S., et al. "Cleaner fish (*Labroides dimidiatus*) recognise familiar clients." *Animal Cognition* 5 (2002): 139-145.
18. Ibid.
19. Bshary, R. "The cleaner fish market." In *Economics in Nature: Social Dilemmas, Mate Choice and Biological Markets*, edited by R. Noë et al., pp. 146-172. Cambridge, U.K.: Cambridge University Press, 2001.
20. Bshary, R., and A. S. Grutter. "Asymmetric cheating opportunities and partner control in a cleaner fish mutualism." *Animal Behavior* 63 (2002): 547-555.
21. Bshary R, Würth M. "Cleaner fish (Labroides dimidiatus) Labroides dimidiatus manipulate client reef fish by providing tactile stimulation." *Proceedings in Biological Science.* 268 (2001):1495-501.
22. Bshary, R. "Machiavellian intelligence in fishes." In *Fish Cognition and Behavior*, edited by C. Brown et al., pp. 223-242. Oxford: Wiley-Blackwell, 2006.
23. Bshary, R., and A. D'Souza. "Cooperation in communication networks: Indirect reciprocity in interactions between cleaner fish and client reef fish." *Communication Networks*, edited by P. K. McGregor, pp. 521-539. Cambridge and New York: Cambridge University Press, 2005.
24. Schuster, R., and A. Perelberg. "Why cooperate? An economic perspective is not enough." *Behavioral Processes* 66 (2004): 261–77.
25. Taborsky, M., and D. Limberger. "Helpers in fish." *Behavioral Ecology and Sociobiology* 8 (1981): 143–145.
26. Wilkinson, G. S. "Communal nursing in the evening bat, *Nycticeius humeralis*." *Behavioral Ecology and Sociobiology* 31 (1992): 225-235.
27. Packer, C., et al. "A comparative analysis of non-offspring nursing." *Animal Behaviour* 43 (1992): 265-281.
28. Kunz, T. H., et al. "Alloparental care: Helper-assisted birth in the Rodrigues fruit bat *Pteropus rodricensis* (Chiroptera: Pteropodidae)." *Journal of Zoology* 232 (1994): 691-700.

29. Dieterlin, F. "Geburi und Geburtshilfe bei der Stachelmaus." *Zeitschrift fur Tierpsychologie* 19 (1962): 191-222.
30. Tavolga, M. C., and F. S. Essapian. "The behavior of the bottlenosed dolphin (*Tursiops truncatus*): Mating, pregnancy, parturition and mother-infant behavior." *Zoologica* 42 (1957): 11-31.
31. Whitehead, H. *Sperm Whales: Social Evolution in the Ocean*. Chicago: University of Chicago Press, 2003.
32. Ehrlich, P., et al. *The Birder's Handbook: A Field Guide to the Natural History of North American Birds*. New York: Simon & Schuster, 1988.
33. DuVal, E. H. "Adaptive advantages of cooperative courtship in the lance-tailed manakin (*Chiroxiphia lanceolata*)." PhD thesis, University of California, Berkeley, 2005.
34. Weidt, A., et al. "Not only mate choice matters: Fitness consequences of social partner choice in female house mice." *Animal Behaviour* 75 (2008): 801-808.
35. Gilchrist, J. S. "Pup escorting in the communal breeding banded mongoose: Behaviour, benefits, and maintenance." *Behavioral Ecology* 15 (2004): 952-960.
36. Corning, *Nature's Magic*.
37. Brotherton, P. N. M., et al. "Offspring food allocation by parents and helpers in a cooperative mammal." *Behavioral Ecology* 12 (2001): 590-599.
38. Clark, R. W. "Kin recognition in rattlesnakes." *Biology Letters* 271 (2004): S243–S245.
39. Huang, W. "Parental care in the long-tailed skink, *Mabuya longicaudata*, on a tropical Asian island." *Animal Behaviour* 72 (2006): 791-795.
40. MacKenzie, D. "Eels and groupers hunt better together." *NewScientist.com*. December 5, 2006.
41. Alfieri, M. S., and L. A. Dugatkin LA. "Cooperation and cognition in fishes." In *Fish Cognition and Behavior*, edited by C. Brown et al., pp. 203-222. Oxford: Wiley-Blackwell, 2006.
42. British Broadcasting Corporation. *The Life of Mammals*. 2002.
43. De Waal, F. B. M. *Our Inner Ape: A Leading Primatologist Explains Why We Are Who We Are*. New York: Riverhead, 2005.
44. Melis, A. P., et al. "Chimpanzees recruit the best collaborators." *Science* 311 (2006): 1297-1300.
45. Bshary, R. "Machiavellian intelligence in fishes." In *Fish Cognition and Behavior*, edited by C. Brown et al., pp. 223-242. Oxford: Wiley-Blackwell, 2006.
46. Cronin, K. A., and C. T. Snowdon. "The effects of unequal reward distributions on cooperative problem solving by cottontop tamarins, *Saguinus oedipus*." *Animal Behaviour* 75 (2008): 245-257.

47. Rutte, C., and M. Taborsky. "Generalized reciprocity in rats." *PLoS Biology* 5 (2007): e196.
48. Höldobler, B., and E. O. Wilson. *The Superorganism: The Beauty, Elegance, and Strangeness of Insect Societies.* New York: W. W. Norton, 2009.
49. Moss, C. *Portraits in the Wild: Animal Behavior in East Africa.* 2nd ed. Chicago: University of Chicago Press, 1982.
50. Ibid.

Chapter Eight

1. Sapolsky, R. M. *A Primate's Memoir: A Neuroscientist's Unconventional Life Among the Baboons.* New York: Scribner, 2001. See pp. 238-239.
2. Maynard Smith, J., and G. Price. "The logic of animal conflict." *Nature* 246 (1973): 15-18.
3. Attenborough, D. *The Trails of Life: A Natural History of Animal Behavior.* Boston: Little, Brown, 1991.
4. Angier, N. "Who is the walrus?" *New York Times.* May 20, 2008.
5. Moss, C. *Portraits in the Wild: Animal Behavior in East Africa.* 2nd ed. Chicago: University of Chicago Press, 1982.
6. Steinhart, *The Company of Wolves*, New York: Vintage, 1996. p. 114.
7. Moss, *Portraits in the Wild.*
8. Koyama, N. F., and R. I. M. Dunbar. "Anticipation of conflict by chimpanzees." *Primates* 37 (1996): 79-86.
9. De Waal, F. B. M. *Our Inner Ape: A Leading Primatologist Explains Why We Are Who We Are.* New York: Riverhead, 2005.
10. Fraser, O. N., et al. "Stress reduction through consolation in chimpanzees." *Proceedings of the National Academy of Sciences (USA)* 105 (2008): 8557-8562.
11. Schino, G. "Beyond the primates: Expanding the reconciliation horizon." In *Natural Conflict Resolution*, edited by F. Aureli and F. B. M. de Waal, pp. 225-241. Berkeley: University of California Press, 2000.
12. Bekoff, M., and J. Pierce. *Wild Justice: The Moral Lives of Animals.* Chicago: University of Chicago Press, 2009.
13. Brosnan, S. F., et al. "Tolerance for inequity may increase with social closeness in chimpanzees." *Proceedings of the Royal Society B: Biological Sciences* 272 (2005): 253-258.
14. Clark, M. S., and N. K. Grote. "Close relationships." In *Handbook of Psychology, Personality and Social Psychology*, vol. 5, edited by T. Millon et al., pp. 447-461. New York: Wiley, 2003.

15. Brosnan, S. F., and F. B. M. de Waal. "Monkeys reject unequal pay." *Nature* 425 (2003): 297-299.
16. Bekoff, M. "Animals' lives matter, so let's stop eating them now." *Daily Camera*, November 25, 2007.
17. Range, F., et al. "Effort and reward: Inequity aversion in domestic dogs?" *Journal of Veterinary Behavior* 4 (2009): 45-46.
18. Egremont, P., and M. Rothschild. "The calculating cormorants." *Biological Journal of the Linnaean Society* 12 (1979): 181-186.
19. Killen, M., and F. B. M. de Waal. "The evolution and development of morality." In *Natural Conflict Resolution*, edited by F. Aureli and F. B. M. de Waal, pp. 352-372. Berkeley: University of California Press, 2000.
20. Warneken, F., and M. Tomasello. "Altruistic helping in human infants and young chimpanzees." *Science* 311 (2006): 1301-1303.
21. Warneken, F., et al. "Spontaneous altruism by chimpanzees and young children." *PLoS Biology* 5 (2007): 1-7.
22. Premack, D., and G. Woodruff. "Chimpanzee problem-solving: A test for comprehension." *Science* 202 (1978): 532-535.
23. Beran, M. J. "Maintenance of self-imposed delay of gratification by four chimpanzees (*Pan troglodytes*) and an orangutan (*Pongo pygmaeus*)." *Journal of General Psychology* 129 (2002): 49-66.
24. Chester, C. *Providence of a Sparrow: Lessons from a Life Gone to the Birds.* New York: Anchor, 2002. See p. 220.
25. Genty, E., and J. J. Roeder. "Self-control: Why should sea lions, *Zalophus californianus*, perform better than primates? *Animal Behaviour* 72 (2006): 1241-1247.
26. Church, R. M. "Emotional reactions of rats to the pain of others." *Journal of Comparative and Physiological Psychology* 52 (1959): 132-134.
27. Cynthia Moss, personal communication, May 2009.
28. McNutt, J., et al. *Running Wild: Dispelling the Myths of the African Wild Dog.* Washington, D.C.: Smithsonian Books, 1997.
29. De Waal, F. B. M. *Our Inner Ape: A Leading Primatologist Explains Why We Are Who We Are.* New York: Riverhead, 2005.
30. Zahn-Waxler, C. et al. "The origins of empathy and altruism." In *Advances in Animal Welfare Science 1984,* edited by M. W. Fox and L. D. Mickley, pp. 21-39. Boston: Martinus Nijhoff, 1984.
31. Langford, D. J., et al. "Social modulation of pain as evidence for empathy in mice." *Science* 312 (2006): 1967-1970.
32. Fouts, R., and S. T. Mills. *Next of Kin: My Conversations with Chimpanzees.* New York: William Morrow, 1997.

33. Hamilton, J. "A voluble visit with two talking apes." National Public Radio, "Morning Edition," July 8, 2006.
34. Parr, L. A. "Cognitive and physiological markers of emotional awareness in chimpanzees (*Pan troglodytes*)." *Animal Cognition* 4 (2001): 223-229.
35. Hamilton, J. "Research shows mice may have feelings too." National Public Radio, "Morning Edition," July 5 2006.
36. De Waal, *Our Inner Ape*. p. 229.
37. De Waal, *Our Inner Ape*.
38. Povinelli, D. J., et al. "Role reversal by rhesus monkeys, but no evidence of empathy." *Animal Behaviour* 44 (1992): 269-281.
39. Galloway, M. "Pachyderm pals." *Washington Post*, February 5, 2007, p. B8.
40. Stahl, J. D. Bos. "The mixed blessings of compulsory group living: Goose flocks impose their own rules." In *Seeking Nature's Limits: Ecologists in the Field*, edited by R. Drent et al., pp. 250-257. Utrecht, The Netherlands: KNNV Publishing, 2005.
41. Prins, H. H. T. *Ecology and Behavior of the African Buffalo: Social Inequality and Decision Making*. London: Chapman & Hall, 1996.
42. Hooper, R. "House-hunting bees behave like a brain." *New Scientist* 197 (2008): 12. The original research was published in *Behavioral Ecology and Sociobiology*.
43. BBC News. "New Zealand dolphin rescues beached whales." British Broadcasting Corporation, March 12, 2008. http://news.bbc.co.uk/2/hi/asia-pacific/7291501.stm.
44. Gareth Patterson, personal communication, April 2008.
45. Dublin, H. T. "Cooperation and reproductive competition among female African elephants." In *Social Behavior of Female Vertebrates*, edited by S. Wasser, pp. 291-315. New York: Academic Press, 1983.

Chapter Nine

1. Dennett, D. C. *Darwin's Dangerous Idea: Evolution and the Meanings of Life*. New York: Simon & Schuster, 1995.
2. Williams, G. C. "Huxley's *Evolution and Ethics* in sociobiological perspective." *Zygon* 23 (1988): 383-407.
3. Dennett, *Darwin's Dangerous Idea*, p. 478; Dennett's emphasis.
4. Dawkins, R. *River Out of Eden*. New York: Basic Books, 1993. See pp. 131-132.

5. Williams, G. C. "Mother Nature Is a Wicked Old Witch." In *Evolutionary Ethics*, edited by M. H. Nitecki and D. V. Nitecki, pp. 217-231. Albany: State University of New York Press, 1993.
6. Williams, "Mother Nature Is a Wicked Old Witch," p. 225.
7. Balcombe, J. P. *Pleasurable Kingdom: Animals and the Nature of Feeling Good.* London and New York: Macmillan, 2006.
8. Bagemihl, B. *Biological Exuberance: Animal Homosexuality and Natural Diversity.* New York: St. Martin's Press, and London: Profile Books, Ltd., 1999.
9. Naskrecki, P. *The Smaller Majority: The Hidden World of the Animals That Dominate the Tropics.* Cambridge, MA: Belknap Press of Harvard University Press, 2005.
10. Colinvaux, P. A. *Why Big, Fierce Animals Are Rare: An Ecologist's Perspective.* Princeton, NJ: Princeton University Press, 1978.
11. Verdolin, J., and C. N. Slobodchikoff. "Vigilance and predation risk in Gunnison's prairie dogs (*Cynomys gunnisoni*)." *Canadian Journal of Zoology* 80 (2002): 1197-1203.
12. Moss, C. *Portraits in the Wild: Animal Behavior in East Africa.* 2nd ed. Chicago: University of Chicago Press, 1982.
13. Howard, L. *Birds as Individuals.* London: Collins, 1952.
14. Sapolsky, R. M. *A Primate's Memoir: A Neuroscientist's Unconventional Life Among the Baboons.* New York: Scribner, 2001. pp. 238-239.
15. Henetz, P. "Wolf's death stirs fears for species' fate." *Salt Lake City Tribune.* April 8, 2008.
16. Ardrey, R. *The Territorial Imperative: A Personal Inquiry into the Animal Origins of Property and Nations.* New York: Atheneum, 1966.
17. Washburn, S. L., and I. DeVore. "The social life of baboons." *Scientific American* 204 (1961): 62-71.
18. Moss, *Portraits in the Wild*, p. 193-194.
19. Strum, S. *Almost Human: A Journey into the World of Baboons.* New York: Random House, 1987.
20. Sapolsky, *A Primate's Memoir*, pp. 238-239.
21. Verhulst, S., and H. M. Salomons. "Why fight? Socially dominant jackdaws, *Corvus monedula,* have low fitness." *Animal Behaviour* 68 (2004): 777-783.
22. Wong, B. B. M. "Superior fighters make mediocre fathers in the Pacific blue-eye fish." *Animal Behaviour* 67 (2004): 583-590.
23. Forsgren, E. "Female sand gobies prefer good fathers over dominant males." *Proceedings of the Royal Society B: Biological Sciences* 264 (1997): 1283–1286.

24. Cheney, D. L., and R. M. Seyfarth. *Baboon Metaphysics: The Evolution of a Social Mind.* Chicago: University of Chicago Press, 2007.
25. Gregory, N. G. *Physiology and Behavior of Animal Suffering.* Oxford: Blackwell, 2004.
26. Anonymous. "Wild horses with wings." *Ecos.* February-March 2006, pp. 18-21.
27. Bruce, H. M. "An exteroceptive block to pregnancy in the mouse." *Nature* 184 (1959): 105.
28. Latham, N., and G. Mason. "From house mouse to mouse house: The behavioural biology of free-living *Mus musculus* and its implications in the laboratory." *Applied Animal Behaviour Science* 86 (2004): 261-289.
29. Lamey, T. C., and D. W. Mock. "The role of brood size in regulating aggression." *American Naturalist* 138(4) (1991): 1015-1026.
30. Price, P. W. *Evolutionary Biology of Parasites.* Princeton, NJ: Princeton University Press, 1980.
31. Charles, H., and J. Godfray. "Parasitoids." *Current Biology Magazine* 14 (2004): R456.
32. Treat, A. E. *Moths of Mites and Butterflies.* Ithaca, NY: Cornell University Press, 1975.
33. Dunn, D. W., et al. "A Role for Parasites in Stabilising the Fig-Pollinator Mutualism." *PLoS Biol* 6 (2008): e59 doi:10.1371/journal.pbio.0060059.
34. Fellous, S., and L. Salvaudon. "How can your parasites become your allies?" *Trends in Parasitology* 25 (2009): 62-66.
35. Janzen, D. H., ed. *Costa Rican Natural History.* Chicago: University of Chicago Press, 1983.
36. Mayr, E. *What Evolution Is.* New York: Basic Books, 2001.
37. Moss, *Portraits in the Wild.*
38. Ibid.
39. Garde, E., et al. "Age-specific growth and remarkable longevity in narwhals (*Monodon monoceros*) from West Greenland as estimated by aspartic acid racemization." *Journal of Mammalogy* 88 (2007): 49-58.
40. Laidlaw, R. *Wild Animals in Captivity.* Markham, Canada: Fitzhenry & Whiteside, 2008.
41. Clubb, R., et al. "Compromised survivorship in zoo elephants." *Science* 322 (2008): 1949.
42. Parker, K. L., et al. "Energy and protein balance of free-ranging black-tailed deer in a natural forest environment." *Wildlife Monograph* 143 (1999): 1–48.

43. Gould, J. L., and C. G. Gould. *Animal Architects: Building and the Evolution of Intelligence.* Chicago: University of Chicago Press, 2007.
44. Camus, A. *Notebooks 1951–1959.* Translated from the French, with an introduction and afterword by Ryan Bloom. Chicago: Ivan R. Dee, 2008.

Chapter Ten

1. Bergman, J. "Mankind: The Pinnacle of God's Creation." Institute for Creation Research, n.d. http://www.icr.org/article/238/.
2. Hewer, C. T. R. *Understanding Islam: An Introduction.* Minneapolis, MN: Fortress Press, 2006.
3. Gould, S. J. *Life's Grandeur: The Spread of Excellence from Plato to Darwin.* London: Jonathan Cape, 1996.
4. Malamud, R. "Life as we know it." *Chronicle of Higher Education* 54 (July 25, 2008): B7.
5. Bakewell, M. A., et al. "More genes underwent positive selection in chimpanzee evolution than in human evolution." *Proceedings of the National Academy of Sciences (USA)*104 (2007): 7489-7494.
6. Allen, J. R., et al. "Biological effects of polychlorinated biphenyls and triphenyls on the subhuman primate." *Environmental Research* 6 (1973): 344-354.
7. Bender, D., and B. Leone. *Violence in the Media.* Farmington Hills, MI: Greenhaven Press, 1995.
8. Sacks, D. *Encyclopedia of the Ancient Greek World.* New York: Facts On File, 1995.
9. White, M. *Selected Death Tolls for Wars, Massacres and Atrocities Before the 20th Century.* Web site: http://users.erols.com/mwhite28/warstat0.htm#20worst.
10. McCarthy, C., ed. *Solutions to Violence.* Washington, D.C.: Center for Teaching Peace, 1990.
11. Lewison, R., and W. Oliver. *Hippopotamus amphibius.* In *2007 IUCN Red List of Threatened Species.* International Union for Conservation of Nature and Natural Resources, 2007. www.iucnredlist.org
12. Cosgrove-Mather, B. "Poaching devastates Congo's hippos." CBS News, November 10, 2005. http://www.cbsnews.com/stories/2005/11/10/tech/main1035887.shtml.
13. Eilperin, J. 2007. "Fish story's new reality: Man bites shark." *Washington Post*, June 19, 2007.
14. Nishiwaki, M. "Ryukyuan whaling in 1961." Scientific Report from the Whales Research Institute Tokyo, vol. 16 (1962): 19-28.
15. Williams, E. E. and M. DeMello. *Why Animals Matter.* Amherst, New York: Prometheus, 2007.

16. Associated Press. "South Africa will kill elephants to control population, animal rights activists threaten boycotts." Associated Press, February 26, 2008.
17. Moss, C. *Portraits in the Wild: Animal Behavior in East Africa.* 2nd ed. Chicago: University of Chicago Press, 1982.
18. Eiseley, L. *The Unexpected Universe.* New York: Harcourt Brace Jovanovich, 1969.
19. Mishna, Sanhedrin. Per Bartlett's Quotations.
20. Gourlay, C. "It's Hen Girl, the superhero swooping in to rescue battery chickens." *Sunday Times* (U.K.), June 29, 2008.
21. Taylor, K., et al. "Estimates for worldwide laboratory animal use in 2005." *Alternatives to Laboratory Animals: ATLA* 36 (3) (2008): 327-342.
22. Moses, H., et al. "Financial anatomy of biomedical research." *JAMA* 294 (2005): 1333-1342.
23. NIMH (National Institute of Mental Health). "Nobelist Discovers Antidepressant Protein in Mouse Brain. Press Release." January 06, 2006.
24. Diamond, J. *Guns, Germs, and Steel: The Fates of Human Societies.* New York: W. W. Norton, 1997.
25. National Public Radio. "The End of the Bullfight (or Not So Much?)" "Blog of the Nation," blog of "Talk of the Nation," National Public Radio, June 5, 2008. http://www.npr.org/blogs/talk/2008/06/the_end_of_the_bullfight_or_no.html.
26. Camus, A. *Notebooks 1951–1959.* Translated from the French, with an introduction and afterword by Ryan Bloom. Chicago: Ivan R. Dee, 2008.
27. Osthaus, B., et al. "Dogs (*Canis lupus familiaris*) fail to show understanding of means-end connections in a string-pulling task." *Animal Cognition* 8 (2005): 37-47.
28. Call, J. "The use of social information in chimpanzees and dogs." In *Comparative Vertebrate Cognition: Are Primates Superior to Non-Primates?* edited by L. J. Rogers and G. Kaplan, pp. 263-286. New York: Kluwer Academic, 2003.
29. "Can Dogs Smell Cancer?" *ScienceDaily.* January 6, 2006. http://www.sciencedaily.com/releases/2006/01/060106002944.htm.
30. Eldridge, N. "The sixth extinction." ActionBioScience.org, June 2001. http://www.actionbioscience.org/newfrontiers/eldredge2.html.
31. Weisman, A. *The World Without Us.* New York: Thomas Dunne Books, 2007.
32. Mowat, F. *Sea of Slaughter.* Toronto: McClelland & Stewart, and Boston: Atlantic Monthly Press, 1984.
33. Diamond, J. *Collapse: How Societies Choose to Fail or Succeed.* New York: Viking, 2005.

34. Glavin, T. *The Sixth Extinction: Journey Among the Lost and Left Behind.* New York: Thomas Dunne Books, 2006.
35. Cave, R. "When a whale fall falls." Letter to the editor. *New Scientist* 199 (2008): 20.
36. Glavin, T. *The Sixth Extinction.*
37. Thomas, C. D., et al. "Extinction risk from climate change." *Nature* 427 (2004): 145-148.
38. Morales, A. "UN warns of extinction, flooding from global warming." Bloomberg News, April 6, 2007. http://www.bloomberg.com/apps/news?pid=20601085&sid=aBQ_pcXl4cVo&refer=europe.
39. Parmesan, C. "Ecological and evolutionary responses to recent climate change." *Annual Review of Ecology, Evolution, and Systematics* 37 (2006): 637-669.
40. Pounds, J. A., et al. "Widespread amphibian extinctions from epidemic disease driven by global warming." *Nature* 439 (2006): 161-167.

Chapter Eleven

1. Daly, H. E., and K. N. Townsend. *Valuing the Earth: Economics, Ecology, Ethics.* Cambridge, MA: MIT Press, 1993.
2. Cloudsley-Thompson, J. L. *Tooth and Claw: Defensive Strategies in the Animal World.* London: J. M. Dent & Sons, 1980.
3. Primack, R. B. *Essentials of Conservation Biology.* 4th ed. Sunderland, MA: Sinauer Associates, 2006. See "Habitat Destruction," pp. 177-188.
4. Geist, H. J., and E. E. Lambin. "Proximate causes and underlying driving forces of tropical deforestation." *BioScience* 52 (2) (2002): 143-150.
5. American Heart Association. Statistical Fact Sheet—Populations. 2008 Update. http://www.americanheart.org/downloadable/heart/1201543457735FS06INT08.pdf.
6. Ehrlich, P. R., and A. H. Ehrlich. *The Dominant Animal: Human Evolution and the Environment.* Washington, D.C.: Island Press, 2008.
7. Food and Agriculture Organization of the United Nations (FAO). "Livestock's long shadow: Environmental issues and options." Rome: Food and Agriculture Organization, 2006. FAO.org.
8. Council for Agricultural Science and Technology (CAST). "Animal Agriculture and Global Food Supply." Ames, IA: Council for Agricultural Science and Technology, 1999. See tables 4.17-4.24.
9. De Rothschild, D. *Live Earth Global Warming Survival Handbook: 77 Essential Skills to Stop Climate Change.* New York: Rodale, 2007.

10. Vince, G. "Earth 2099: How to survive the coming century." *New Scientist,* February 25, 2009.
11. Stein, R. "Daily red meat raises chances of dying early." *Washington Post,* March 24, 2009.
12. Black, R. "Shun meat, says UN climate chief." British Broadcasting Corporation, September 7, 2008. http://news.bbc.co.uk/2/hi/science/nature/7600005.stm.
13. Pigott, D. C. "Foodborne illness." *Emergency Medicine Clinics of North America* 26 (2008): 475-497.
14. Delgardo, C., et al. "Livestock to 2020, the next food revolution." Food, Agriculture and the Environment Discussion Paper 28, prepared for the International Food Policy Research Institute; the Food and Agriculture Organization of the United Nations (FAO); and the International Livestock Research Institute, May 1999. PDF available at www.ifpri.org/2020/dp/dp28.pdf.
15. Physicians Committee for Responsible Medicine. "Angry doctors use 1,000 tomatoes to spell out salmonella source for FDA: 'It's the meat, stupid!' " News release, July 9, 2008: http://www.pcrm.org/news/release080709.html.
16. Greger, M. *Bird Flu: A Virus of Our Own Hatching.* New York: Lantern, 2006.
17. Kennedy, M. "Bird flu could kill millions: Global pandemic warning from WHO. 'We're not crying wolf. There is a wolf. We just don't know when it's coming.' " *Montreal Gazette,* March 9, 2005.
18. Ibid.
19. Animals and Society Institute, 2009: www.animalsandsociety.org.
20. Pancevski, B. "New Swiss law protects rights of 'social' animals." *Times Online* (U.K.), April 26, 2008. http://www.timesonline.co.uk/tol/news/world/europe/article3818457.ece.
21. Lund, V., et al. "Expanding the moral circle: Farmed fish as objects of moral concern." *Diseases of Aquatic Organisms* 75 (2007): 109-118.
22. Kimmelman, M. "Bullfighting is dead! Long live the bullfight!" *New York Times,* June 1, 2008.
23. Tarvainen, S. "Pressure mounts to ban bullfights in Spain's Catalonia." Deutsche Press-Agentur, May 19, 2009.
24. Crary, D. "Number of U.S. hunters steadily declines, worrying state wildlife agencies." *USA Today,* September 2, 2007.
25. Schiermeier, Q. "Primate work faces German veto." *Nature* 446 (2007): 955.
26. Abbott, A. "Swiss court bans work on macaque brains." *Nature* 453 (2008): 833.

27. Perry, J., and M. Jacoby. "These little pigs get special care from Norwegians." *Wall Street Journal*, August 6, 2007. http://www.meatintelligence.com/news/articles/55245.shtml.
28. UPC (United Poultry Concerns). nd. "Forced Molting." http://www.upc-online.org/molting/.
29. Party for the Animals (PftA). 2009. http://www.partyfortheanimals.info/content/view/300.
30. Diamond, J. *Collapse: How Societies Choose to Fail or Succeed.* New York: Viking Penguin, 2005.
31. Kellert, S., and A. Felthous. "Childhood cruelty toward animals among criminals and noncriminals." *Human Relations* 38 (1985): 1113-1129.
32. Merz-Perez, L., et al. "Childhood cruelty to animals and subsequent violence against humans." *International Journal of Offender Therapy and Comparative Criminology* 45 (5) (2001): 556-573.
33. Fitzgerald, A. J., et al. "Slaughterhouses and increased crime rates." *Organization & Environment* 22 (2009): 158-184.
34. Salt, H. *Animals' Rights: Considered in Relation to Social Progress.* 1892. Reprinted with a preface by Peter Singer. Clarks Summit, PA: Society for Animal Rights, 1980.
35. Monsaingeon, B. *Glenn Gould: Hereafter.* A film by Bruno Monsaingeon. Idéale Audience, 2005.
36. Stowe, H. B. *Uncle Tom's Cabin.* New York, Simon & Schuster (1852, 1963).

index

abnormal behavior 52
activism 200–204
affection 50, 124
African wild dog
  hunting 118–119
  midwifery 112
  restraint 130
aggression 122
Aghion, Philippe 188
agouti 122
alarm calls 72–73, 88–89, 98–99, 109
altruism 105–107, 121, 127–128, 136
Amboseli Elephant Research Project 54
Angier, Natalie 123
animal protection
  advancement of 195–200
  European Union 196
  laws 195–197
animals
  capacities 4
  challenges faced by 15
  different from humans 14, 15
  human treatment 4–5, 14, 141–142
  liberation of 7–8
  media portrayal 148–149
  perceptions 16–30
  psychiatry industry 58
  sensory systems 15
  sociability 78, 103–120
Animals & Society Institute 195
Animal Welfare Act (U.S.) 42
anthropocentrism 164–165, 188
anticipation 16, 69
ants
  agriculture 118
  mutualism 99, 105
  cooperation 118
aphid 100, 105
Ardrey, Robert 152
Aristotle 77, 164
attention 53–54, 68
attribution 63–65, 67
audience effects 41–42, 69–70
awareness 41–42, 61–78
Ayumu (chimpanzee) 32–33
babbler (bird)
  play 53
  self-handicapping 130
  social tensions 53

baboon
  courage 121
  friendships 47
  grief 47–48
  infanticide 154
  misconceptions of 152–153
  sharing 106
  solidarity 118
  stress 47–48
  survival with handicap 151
babysitting 112–113
Backensto, Stacia 70
bacteria 104–105, 156, 194
Bagemihl, Bruce 147
Barnard, Neal 194
Barro Colorado Island 23
bats 9–13
  communal nursing 112
  diversity 9
  eavesdropping 20

bats—*Continued*
  echolocation control 100
  feeding behavior 20
  hearing 18–19
  call signatures 11
  flight 11–13
  individual recognition 10–11
  mother-pup reunions 9–11
  navigation 12
  perceptions 18–19, 22
  predation 12
  red bat 20
  midwifery 112
  success of 9
  ultrasound 19
  sharing 106–107
Battery Hen Welfare Trust 174
beauty 3
beaver
  flexible behavior 29
  persecution of 75
  planning 75
  problem-solving 75–76
  tool-use 75–76
Beecher, Henry Ward 31
bees
  democracy 135
  language 87
beetle, bombardier 28
Bekoff, Marc 17, 24, 45, 46, 49
bereavement (see grief)
Bergman, Jerry 164
biophilia 204
birds
  brains 66
  communication 100–101
  cooperative breeding 113
  cooperative hunting 115
  deception 72–74
  extinction 181–182
  individual recognition 96
  memory 34–35, 96
  perceptions 21–22
  siblicide 155
bison 56–57
Boggs, Lesley 130
bonobo
  empathy 132
  problem-solving 94
Boudreaux, Donald 187
bowerbird 161
Bradshaw, Gay 58
brains
  and emotion 125, 131
  and sentience 16–17, 180
  as disadvantageous 31–32
  evolution of 66, 179–180
British Union for the Abolition of Vivisection 174
Bronstein, Judith 103
Broom, Donald 17
Bruce Effect 155
Bruce, Hilda Margaret 155
Bshary, Redouan 115
Bugnyar, Thomas 74
bullfighting
  arguments for and against 178
  waning public support 196–197
Burghardt, Gordon 39
Burke, Joanna 55
butterflies 75

Camus, Albert 162, 179
cannibalism 155
cape buffalo
  democracy 135
  solidarity 119
captivity
  effect on behavior 155
  longevity in 160
capuchin monkey
  sense of fairness 125–126
  sharing 106
carnivores
  cooperation 111
  humans as 191–192
  virtue 126
carrying capacity 161
cat, domestic
  concern for others 131
  emotions 45–46, 49, 58
  reconciliation 125
  stress 57–58
cattle
  magnetic orientation 21
  emotions 60
Cave, Rachel 183
cheating 109–110

cheetah 160
Cheney, Dick 70
Cheney, Dorothy 66, 73, 154
chickadee
  communication 89, 98
  mobbing 89
chickens
  aesthetic preferences 3
  alarm calls 73, 88–89
  as sentient individuals 4, 171–172
  cannibalism 155
  communication 4
  perceptions 3
  protection of 174
  referential calls 87–88
chimpanzee
  alliances 72
  altruism 127–128
  attribution 65
  awareness 93
  collaboration 116
  consolation 125
  deception 64, 71
  emotional awareness 132–133
  emotional fever 48–49
  empathy 131, 133
  evolution 166
  gestures 65, 92
  gratitude 50
  humor 71
  infanticide 154
  memory 32–33, 59, 180
  moral development 127
  peacemaking 124
  perceptiveness 95
  problem-solving 94
  recognition of others 95
  reconciliation 125
  restraint 128
  sense of fairness 125
  Stroop effect 43
  symbolic communication 92–93
  theory of mind 92
China 191
cities, impact on birdsong 22
Clayton, Nicky 34
cleaner fish
  as mutualists 105, 109–111
  cheating 109–110
  client fish relations 109–110
    role of pleasure 110
  image-scoring 110
  punishment 110
climate change 5, 183, 190, 192–194
Coetzer, William 169
cognition (see intelligence)
Colinvaux, Paul 149
colonialism 177–178
commensalism 156
communication 4, 21, 25, 49–50, 83–86, 89, 92–94, 99–101, 115
competition 103, 107–108, 114
conflict
  avoidance 122–125
  resolution 152
Conradt, Larissa 135
consolation 125
contra-freeloading 37
cooperation 103–120
cormorant 126–127
Corning, Peter 107
corticosterone 18
courtship 113
cowbird 158
coyotes 172
creeper, brown 22
crocodile
  homing 23
  nest guarding 114
culture 86, 99
Cummings, Teresa 88
cuttlefish
  camouflage 84
  communication 83–84

Daly, Herman 186
dairy production 170–171
Darwin, Charles 46, 108
Davis, Karen 174
Dawkins, Richard 144–145
death-feigning 73
deception 28–29, 64, 71–75
  as non-cognitive 71
deer
  conflict-resolution 152
  democracy 135
  magnetic orientation 21
deer mice 4

deference 133–134
degu 76
democracy 135
Dennett, Daniel 144–145
Descartes, René 27, 44, 171
Devore, Irven 153
de Waal, Frans 50, 116, 133
dialects 86, 90
Diamond, Jared 178, 182
dikdik (antelope)
  antipredator responses 43
  survivorship 159
Disney, Walt 143
dissection
  frog 7
  laws 195
dogs
  attention 53–54
  awareness 42, 65, 68
  concern for others 131
  deference 134
  imitation 68
  intelligence 180
  perceptions 18, 21, 180
  play 24
  sense of fairness 126
dolphin
  altruism 136
  cooperation with birds 115
  delayed gratification 33
  individual labeling by 97
  infanticide 154
  midwifery 112
  reconciliation 125
  teaching 34
  planning 33–34
dominance
  as suboptimal strategy 154
  overemphasis on 153–154
Douglas-Hamilton, Iain 54
downer cattle 198–199
Dr Hadwen Trust 174
drongo 98
drug testing 176
ducks 172
Dutch party for the Animals 199
DuVal, Emily 113
Dzeidzic, Walter 168
eavesdropping 20
echolocation 20, 77, 100, 179
Eiseley, Loren 173
electric communication 21, 100
elephant
  altruism 136–137
  communication 85
  conflict with humans 172–173
  culling 55, 59, 172–173
  deference 134
  fights 123–124
  in captivity 160
  interest in other elephants 54
  longevity 160
  midwifery 112
  mortality 159
  perception 85
  persecution 56
  post-traumatic stress disorder 58
  regret 55
  self-handicapping 130
  surprise 63
  urine studies with 62–63
Elephant Sanctuary, Tennessee 55
elk 124
emotions 45–60
  communication 49–50
  evolution of 46
  human compared to animal 60
  implications 53
  survival value 46–47
emotional fever 48
empathy 130–133
endosymbiotic theory 104
energetics of ecosystems 149
erotic behavior 147
Evans, Chris 73
evening bat 112
evolution
  lacking purpose 164
  of cooperation 104–105, 113–122
  of sentience 13, 17
  process of 81, 146, 165–166
evolutionarily stable strategy (ESS) 122
extinction 178, 181–183
  human-caused 181–183
  rates of 181, 183
extravagance in nature 161

factory farming 197–198
fairness 125–127
falcon 151
Fauna Foundation 59
feelings, privacy of 16
fetal calf serum 177
Fields, Bill 132
fisheries
  bycatch 182
  catch statistics 182
  pirate fishing 182
fishes
  audience effects 69
  awareness 41–42
  cognition 110
  communication 86, 100, 115
  cooperation 11–112, 115, 116
  discrimination 69
  dominance 154
  evolution of 165–166
  human prejudice against 39, 166
  non-lethal study methods 169
  perceptions 22, 40–42
  predator inspection 41, 116
  protection of 42, 196
  sentience 42
Frank, Anne 185, 201
Freud, Sigmund 127
Food & Agriculture Organization 13–14, 191, 195
food-borne illness 194
food chains 149, 192
frog
  dissection of 7
  communication 84–85
  extinction 183
  semaphoring 84
  triangulation by 85
  ventriloquy 85
fruit fly 8
fur industry 14

Galloway, Marie 134
Gardner, Allen and Beatrice 87, 132
gaze monitoring 63, 72, 92, 135
gecko 99–100
genocide 168
gestures 65, 83
giraffe 159, 161
Glavin, Terry 183
glucocorticoid 46
goat 17, 125
Godard, Renee 96
Goddard, John 160
goldfish 77
Golding, William 127
Goodall, Jane 95, 106
goose, Canada 108
Gore, Al 193
gorilla
  deception 64, 71
  infanticide 154
  size 192
Gould, Glenn 203
Gould, Stephen jay 44, 164
Grandin, Temple 78
gratitude 50
great ape
  gaze-following 63
  gestures 83
  sign language 87
  theory of mind 64–65
Great barrier reef 105
great tit 22
  awareness 96
  calling by 101
  deception 72
  perception 90
Green Revolution 189
Greger, Michael 194
grief 47, 57
ground squirrel 77
Griffin, Donald 29
grooming 47–48, 64, 112, 124–125, 153
growth
  and habitat loss 189
  economics and 186–189
  of human population 186, 190
    neglect of 190
    problems with 189–191
  role of innovation 188, 189
guinea pig 123
gull
  awareness 61–62
  conflict resolution 122

guppy
  cooperation 116
  discrimination by 41

Hamilton, William 105
handicap
  occurrence in nature 150–152
  survival with 150–152
hawk, red-tailed 12
Heinrich, Bernd 70, 180
helping (see cooperation)
Helvétius 7
herring 86
Hetherington, Thomas 84
Hewer, CTR 164
hippopotamus
  dangerous reputation 167–168
  persecution 168
  population declines 168
Hobbes, Thomas 145
Hölldobler, Bert 118
homing 23–24
honeybees (see bees)
hornbill 98–99
horse 154, 155
horseshoe bat 18–19
Howard, Len 89–90, 96, 151
Howarth, Jane 174
Howell, Mike 169
howler monkey 23
humans
  capacity to change 200–202
  compared to chimps 128
  criminality 202
  cruelty 170, 202
  emotional fever 48
  evolution 166
  failure of empathy 132
  impact on animals 4, 14
  infanticide 154
  mortality 191, 194
  perceptions 94
  perception of nature 143–144
  relationship to animals 13–14, 136, 137, 141, 147–148, 163–166, 185–204
  social advances 14
  superstitions 148
human children
  attribution 67
  empathy 131
  moral development 127
Humane Society of the United States 14, 194, 195, 198–199
hummingbird, rufus 35–36
Humphrey, Nicholas 30
hunting 14, 56, 91–92
  bison 56–57
  declining popularity 197
hyena
  deception 72
  infanticide 154
  longevity 160
  media portrayal 148
  persecution of 173
  reconciliation 125

Imire Safari Ranch 56
impala 74
individual recognition 10–11, 114, 95
influenza 194–195
information sharing 107
insects
  communication with vertebrate 99
  cooperation 111
  democracy 135
  buffer against parasitism 158
  social, success of 119
instinct 27–29
intellicentrism 31, 180
intelligence 31–44
  moral relevance 179–180
  scientific interest 4
International Association for Animal Trauma and Recovery 58
International Union for Conservation of Nature and Natural Resources (IUCN) 168

jackdaw
  dominance effects 154
  sharing 106
  survival with handicap 150–151
jay, scrub
  caching behavior 34–35
  memory 34–35

Kaminski, Juliane 68
kangaroo rat 32
Kanzi (bonobo) 132
Kavanau, Lee 25–27
killdeer 28–29
killing
  by animals 145, 148, 168
  by humans 4, 14, 132, 167–169, 172–173, 175–176, 189, 191, 198
kin selection 105, 107, 113
Klimt, Gustav 185
Klingel, Hans and Ute 97, 119
knifefish 21, 100
Kotrschal, Kurt 74
Kraus, Karl 128
Kruger National Park 119, 173
Kruuk, Hans 160

labeling 97
Ladygina-Kohts, Nadie 130
Laidlaw, Rob 160
language 87–95
  and intelligence 94
  and sentience 94–95
  debasing animals 166
  defining 87
  dulling effect on perceptions 93–94
  referential calls 87–88
laughter 71
learning 26–27, 28, 34–36, 38–41
Ledgard, J. M. 143
lemur
  call recognition by 98
  infanticide 154
leopard 160
Levinson, David 202
Lewiston, Rebecca 168
  lifeas worth living 159, 162
  irrelevance of reproduction 159
life expectancy 160
Lindquist, Erik 84
lion
  infanticide 154
  persecution of 173
  reconciliation 125
  sociability 110
  mortality 160
Little, Rob 173
lizard 48, 99–100, 148
Lockwood, Randall 124
locust 8
longevity 160

macaque
  deference 133
  empathy 133
  self-control 129
  tolerance 133
magpie 66
Malthus, Thomas 189
manatee 24
mangabey 23
manikin 113
Marder, Amy 58
Margulis, Lynn 104
marmoset
  cooperation 117
  midwifery 112
  sharing 117
Marzluff, John 70
Matsuzawa, Tetsuro 32–33
Maynard Smith, John 122
Mayr, Ernst 158
McCarthy, Colman 167
McComb, Karen 54
McNutt, John 130
meat consumption
  China 191
  environmental impact 191, 193–194
  rates 195
  unsustainability of 192
    neglect of 192–193
Mech, David 70
meerkat 114
memory 26, 33, 34–37
Menzel, Emil 93
Mereschkowsky, Konstantin 104
metacognition 67
Mexican free-tailed bat
    call signatures 11
    flight 11–13
    individual recognition 10–11
    mother-pup reunions 9–11
    navigation 12
    predation 12

mice
  activity levels 26–27
  communal nursing 113–114
  communication 25
  compared to primates 26
  drug self-administration by 53
  empathy 131
  infanticide 154
  learning 26–27, 28
  memory 26
  midwifery 112
  need for control 26–27
  pain 131–132
  paternal care 155
  perceptions 25–27
  senses 25–26
  transgenic 176
  vivisection 175–177
midwifery 112
might-makes-right 177–178
mimicry
  birds 98
  butterflies 75
mirror neurons 131
mirth 71
mites 157
mitochondria 104
mongoose
  cooperation 114
  reconciliation 125
monkeys
  abnormal behavior 52
  attention 68
  communication 88
  conflict resolution 125
  deception 71–73
  deference 133
  fairness 125–126
  memory 23
  metacognition 67
  perception 77
  self-awareness 66
  sharing 106–117
  survivorship 150
  vivisection 197
Monty Python 141
morality, evolution of 137
moray eel 115
mortality 159–160
Moskito, James 50
Moss, Cynthia 57, 74, 130, 136, 153, 159
moths
  bat detection 20
  relationship to mite parasite 157
mourning doves
  hunting of 14
  study of 174
Müller, Friedrich Max 87
Munthe, Axel 163
Murphy, Christopher 85
music 40
mutualism 99, 103–105, 156–158

Nairobi National Park 95, 159
naked mole rat
  abilities of 179
  sharing by 106
  tool-use 76–78
narwhal 160
Nash, Ogden 190
Naskrecki, Piotr 148
nature
  cruel and harsh 143–146
    rationalizations for 145–146
  human perceptions of 143–146
Nicastro, Nicholas 49
nightingale 21
noise pollution 86, 100–101
nuthatch 98

optimism 51
orangutan
  problem solving 34
  self-control 128
orca 99
oropendola 158
Orwell, George 15
oryx 152
Osterholm, Michael 194
oxpecker 109

Pacelle, Wayne 199
Pachauri, Rajendra 194
pain 17, 18, 131–132
parasitism
  benefits 157, 158

cowbird 158
detection of 25, 40
host relationship 157
prevalence 156, 158
reproduction 54
parrot
communication with humans 94
language training 94
neurotic behavior 52
problem solving 94
parrotlet, spectacled 97
Pasternak, Kathryn 148
Patterson, Gareth 56, 136
Pauliina, Laurila 134
peacemaking 124–125
Perelberg, Amir 110
personality 51
pessimism 51
People for the Ethical Treatment of Animals (PETA)
Pew Commission 191
Physicians Committee for Responsible Medicine 194
pig
anticipation 69
emotions 49
optimism and pessimism 52
self-handicapping 130
Pixar 103
planthopper 99–100
playback experiments 20, 98, 99, 100
play signals 130
pleasure
as part of life 161
importance of 146–147
in fish 110
science's neglect of 111
pocket gopher 77
polar bear 160
pollination 157
Poplar Spring Animal Sanctuary 88
Posage, Michelle 58
post-traumatic stress disorder 58–59
prairie dog
dialects 90
persecution of 91–92
predation on 150
referential calling 90
predation
difficult lifestyle 159–160
media focus 147–150
success rates 150
predator inspection 41, 77, 116
prejudices 25, 39
Premack, David 94
Pridmore, Ben 32–33
Primatt, Humphrey 16
problem solving 34, 39
Pruetz, Jill 33
psychological trauma 58–60

rabbits
learning 38–39
problem-solving 39
response to chicken calls 89
vivisection 174, 177
raccoon dog 112
Range, Friedericke 126
rat
cooperation 117
drug self-administration by 53
emotional fever 48
infanticide 154
metacognition 67
optimism and pessimism 52
restraint 129–130
sociability 110
success 75
tactile perception 24
virtue 117
rattlesnake
individual recognition 114
prejudices against 114
sociability 114
raven
awareness 61–62
cooperation 108
deception 74
intelligence 180
wariness 70
reciprocal altruism 105–107
reconciliation 125
red bat 20
reefs 105
regret 55
Reichert, Susan 122

Reichmuth, Colleen 24
religion 164, 165
reptiles
  homing 23
  emotion 48
  sociability 114
  cooperative hunting 115
restraint 109, 128–130
rhesus monkey 77
rhinoceros
  crime victims 55, 59
  longevity 160
Richard P. B. 75
Ristau, Carolyn 28
rodents
  abnormal behavior 52
  communal nursing 114
  diversity 9, 27, 75
  fetal resorption 155
  human prejudice against 25
  tool-use 75–77
Rodrigues fruit bat 112
Rollin, Bernard 17
Roper, Tim 135
Royal Ontario Museum 7
Royal Society for the Protection of Birds 172
Rudnai, Judith 95
Rutte, Claudia 118
Rwenzori National Park 136

Sagan, Carl 83
Sagan, Dorion 104
salmon 143–144
Salt, Henry 202
Samburu Reserve, Kenya 54
Sapolsky, Robert 121, 151, 153
Savage-Rumbaugh, Sue and Duane 94
scala naturae 164–165
Schaller, George 160
Schuster, Richard 110
scientific dogma 29–30
seal
  orca recognition by 99
sea lion 129
self-awareness 65–66
self-handicapping 130
selfish genes 105
selfishness 123
sentience 15–18
  animals compared to humans 16–18
  evolution of 13
  experienced by individuals 171–172
  moral significance 13, 180, 203
  relation to intelligence 16
Seyfarth, Robert 66, 73, 154
sharing 106, 117, 123
sharks
  attacks on humans 169
  negative perceptions of 147–148
  persecution of 169
Shaw, George Bernard 175
sheep
  non-random spacing 77
  reconciliation 125
shrikes 72
siblicide
  incidence 155
  barriers to 156
sifaka 98
sign language 87
Silverman, Jerald 17
Simon, Julian 187
Singer, Peter 180
skepticism 37
Skinner, BF 127
Skinner box 46
slaughterhouses 202
Slobodchikoff, Con 90, 107
Smith, Malcolm 136
snake 114, 115, 152
sociability78, 103–120
  advantages of 103
Society for the Prevention of Cruelty to Animals 203
solidarity 111, 118–120
sparrow
  mirror neurons 131
  perceptions 89
  restraint 129
species-centrism 171–172
squirrel
  deception 74–75
  perception 23
  recognition of 95
  reconciliation 125
  vigilance 69
Stachowski, Kathleen 56–57

starling
  awareness 42
  emotions 51
stickleback 40
Stockholm University 3
stoicism 17, 18
stress 17–18, 57–59
Stroop effect 43
Stowe, Harriet Beecher 204
Struhsaker, Thomas 87–88
Strum, Shirley 153
survivorship 159–160
swallow 107
swan 135
swordtail (fish) 41
symbioses 104–105

Taborsky, Michael 117
tamarin 117
teaching 34
technology
  aiding animal study 24
  and population growth 187–188
telepathy 22
Templeton, Christopher 89
Tennyson, Afred Lord
tetra 21
theory of mind 63, 65, 92
time perception 21–23
tit-for-tat 122
Tolle, Eckhart 16
tool-use 65, 75–77, 179
Townsend, Kenneth 186
toxicity testing 175
Trevor, Simon 136
Trivers, Robert 105
turtle 48

umwelt 18–21, 77
United Poultry Concerns 174
unselfish organisms 105
untouchables 163
U.S. Fish and Wildlife Service 152, 172, 197

vampire bat 148
van Lawick, Hugo 95, 118, 134
veganism 201
vegetarianism 188, 192, 201–202, 203
vervet monkey
  deception 73–74
  referential calls 87–88
vigilance 29, 41, 69, 77, 108–109
violence
  human fascination with 147–148, 167
  in human society 202
  in nature 152–153, 158
  link to animal cruelty 202
  on television 167
  link to slaughterhouses 202
virtue 121–138
  evolution 114–115
  rats 117
Virunga National Park 168
vivisection
  behavioral deprivation 37, 52
  calves 177
  efficacy 176
  killing methds 176
  language of 166
  political opposition 197
  number of animals 14, 174
  and animal suffering 17–18, 17–177
vole, meadow 37
von Uexküll, Jakob 18, 21
vulture 73

walrus
  communal nursing 123
  tactile perception 24
  virtue 123
war 167
warbler
  cooperation 113
  expectations 96
  magnetic perception 21
  neighbor recognition 96
  long-term memory 96
Washburn, Sherman 152
Washoe (chimp) 87, 132
wasps 156–157
Watson, John B 127
Webster, John 16
whale
  babysitting 112–113
  communication 85
  culture 86
  dialects 86

whale—*Continued*
gratitude 50
ecosystem importance 182
longevity 160
rescued by dolphin 136
whiskers 24
White, Matthew 167
Wilde, Oscar 128
wildebeest
conflict resolution 124
perceptions 23
Wilkinson, Jerry 106–107
Williams, George C 144–145
Wilson, Ben 86
Wilson, Edward O 118, 181, 204
witnessing effects 18
wolf
conflict resolution 122, 124
intelligence, compared to dog 68
persecution of 170, 172
survival with handicap 151–152
wariness 70
Woodford, Simon 136
woodpecker 185
World Animal Net 195
World Health Organization 191
wren
extravagance 161
plain-tailed 22

Yellowstone National Park 56, 151–152
York University 7

Zahn-Waxler, Carolyn 131
zebra
grief 57
individual recognition 96–97
solidarity 118–119
Zoocheck Canada 160
zoonoses 194
zoos 160